KFS 对 话 建 筑
Architecture Forum

KFS DESIGN IN CHINA IN **20** YEARS
建筑设计在中国的 **20** 年

KFS ARCHITECTS INC. CANADA
PETER GUO HUA FU DESIGN INTERNATIONAL INC. CANADA
加拿大KFS国际建筑师事务所
加拿大傅国华建筑师事务所 编
SELECTED AND CURRENT WORKS FOR 100 PROJECTS
作品精选**100**例

2001－2020

大连理工大学出版社
DALIAN UNIVERSITY OF TECHNOLOGY PRESS

图书在版编目(CIP)数据

KFS对话建筑：建筑设计在中国的20年：汉英对照 /
加拿大KFS国际建筑师事务所，加拿大傅国华建筑师事务
所编. -- 大连： 大连理工大学出版社，2021.10

ISBN 978-7-5685-3135-1

Ⅰ. ①K… Ⅱ. ①加… ②加… Ⅲ. ①建筑设计—作品
集—加拿大—现代 Ⅳ. ①TU206

中国版本图书馆CIP数据核字（2021）第158426号

出版发行：大连理工大学出版社
　　　　　（地址：大连市软件园路80号　邮编：116023）
印　　刷：辽宁新华印务有限公司
幅面尺寸：220mm×300mm
印　　张：21
字　　数：439千字
出版时间：2021年10月第1版
印刷时间：2021年10月第1次印刷
责任编辑：张　泓
责任校对：裘美倩
封面设计：加拿大傅国华建筑师事务所

ISBN 978-7-5685-3135-1
定　　价：328.00元

电　话：0411-84708842
传　真：0411-84701466
邮　购：0411-84708943
E-mail：jzkf@dutp.cn
URL：http://dutp.dlut.edu.cn

本书如有印装质量问题，请与我社发行部联系更换。

20年100例，设计在中国

2020年标志着我们在中国走过了二十年。

城市与社会是整个城市新建筑财富的受益者。从2005年1月1日起，KFS以特有的社会责任感，在上海教育电视台投资创办了一档周播的建筑普及与专业访谈节目——每周六21点40分的《KFS对话建筑》。节目与我们的设计作品一样受到了社会各界的广泛关注和赞誉。我们踏踏实实地走过了20年，坚持建筑师的社会责任心，力求对社会、对城市有所贡献，使城市和建筑不留下遗憾和瑕疵。

在20年的设计中，"始终坚持设计创造价值"是我们一直秉承的宗旨。强调建筑的独创性和唯一性，通过独特的规划建筑构思，为中国的城市发展带来更多的价值。

整整20年过去了，回首看这20年的步伐，KFS仍是那样稳健和从容，力求继续走在行业的前列，要求自己不断地洞察行业形势、引领多元发展。尽管很辛苦，但也很快乐。

未来，KFS希望带给城市与建筑的理念是：多元、复合、以人为本、充满活力。

傅国华 博士
创始人、总裁
2021.1

100 Projects in 20 Years, Design in China

2020 marks our 20th anniversary in China.

The urban and social communities are the beneficiaries of new architectural wealth in the city. As a result of KFS architecture's unique sense of social responsibility, over the years, KFS established a popular weekly professional talk show entitled "KFS Architecture Forum", at 21 : 40 every Saturday on the Shanghai Education Channel from January 1, 2005. The show received positive reviews and interest from different sectors of society since its inception, as did our design work. We have made a 20-year-long journey with the fundamental responsibility that comes with being architects to society, contributing to both the city and its people without compromise.

In 20 years of design practice, our preeminent principle has been "Always adhere to the purpose of creating value by design". The emphasis on originality in architecture and uniqueness in the design process has increased the value of urban development in China.

A full 20 years have passed, while looking at the pace of these 20 years, KFS is still steady and calm, striving to continue to be at the forefront of the industry, furthering research into the industry situation and leading diversified development. KFS continues to overcome hardships with a joyous attitude.

In the future, the concept that KFS hopes to bring to cities and architecture is: diversity, complexity, human focused, and full of vitality.

Peter Fu   Ph.D.
Founder, President
January, 2021

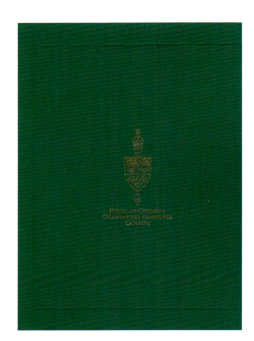

HOUSE OF COMMONS
CHAMBRE DES COMMUNES
CANADA

Ottawa, Ontario
K1A 0A2

September 8, 2017

Dear Dr. Fu:

I am writing to express my sincere gratitude to you for your generous gift to our shared alma mater, McGill University.

As one of the oldest schools of architecture in the country, McGill's distinguished graduates have long shaped the design and culture of Canada and places around the world, each contributing their unique talents and vision to our society. As we celebrate our 150th anniversary of Confederation and Montreal's 375th anniversary, it is all the more important to recognize the significance of our schools in empowering generations of students.

As one of the largest gifts of its kind in Canada, your exemplary support will ensure that the excellence that defines the McGill School of Architecture will continue in the future, creating educational opportunities for students in Canada, China, and around the world. As Canada and China work together to strengthen economic ties, initiatives like this will play a key role in bringing our countries together and in forging new links, new partnerships and new relationships.

Thank you once again for your generous contribution. The legacy of your gift will reverberate for generations to come.

Sincerely,

Dr. Peter Fu

麦吉尔大学校友，加拿大总理特鲁多给傅国华博士的祝贺信　2017.9
McGill Alumnus, Canadian Prime Minister Justin Trudeau's
Letter of Gratitude to Dr. Peter Fu　2017.9

加拿大前总理史蒂芬· 哈珀与傅国华博士 2009
Former Prime Minister of Canada Stephen Harper
and Dr. Peter Fu 2009

加拿大前总理克雷蒂安与傅国华博士 2004
Former Prime Minister of Canada Jean Chrétien
and Dr. Peter Fu 2004

目录　CONTENTS

城市规划和城市设计　URBAN PLANNING & URBAN DESIGN

公共建筑　PUBLIC DESIGN

居住建筑 · RESIDENTIAL DESIGN

室内设计   INTERIOR DESIGN

URBAN PLANNING & URBAN DESIGN

城市规划和城市设计

2010年上海世界博览会申博方案规划
Shanghai World Expo 2010 Bidding Proposal Submission

地点：中国，上海
业主：上海市城市规划管理局
设计范围：规划设计
设计时间：2001

Location: Shanghai, China
Client: Shanghai Urban Planning Administration Bureau
Design scope: Planning Design
Design time: 2001

KFS非常有幸成为上海市人民政府所邀请的七个申博方案设计单位之一，也是唯一一个来自北美的设计单位。

2010年上海世博会的申博方案设计灵感来源于现存的地理环境特征：依托现有黄浦江支流的走势，增加部分人工河道，将展会的主体内容沿着这条圆形的轴线来布置。轴线的中央是大型绿地和休闲设施。由于整个规划区域被黄浦江分隔，因此如何将人流运输到对岸成为本次设计主要考虑的问题之一。

北部扇形区为其他永久性展馆的用地和一个巨大的开放区域。在人工湖半径之外，黄浦江南侧是留给主要国家展馆的用地。圆是整个概念的主题，八个塔式建筑将分布于圆的周围，成为限定圆形空间的"主柱"。作为2010年上海世博会主题的一部分，设计提倡对可二次利用和可循环使用的材料的开发利用。

世博会共开展184天，总参观人次预计达7 000万。为适应高峰时平均每天约70万人次的交通量，该地区的建筑容积率可达4.5左右。设计面临的挑战是：需要寻找一个能在世博会结束后同样对该地区有长期积极作用的规划。一部分结构将成为户外开放空间。圆形人工河之外的塔式建筑将会留作酒店或变成居住建筑。以上都体现了"城市，让生活更美好"这一世博主题。

总用地面积：　　　　　　500 hm²

KFS is very fortunate to be one of the seven design units selected for the bid solicitation by the Shanghai Municipal People's Government, as the only design unit from North America.

The bidding proposal for Shanghai World Expo 2010 was inspired by the existing characteristics of the geographical environment: according to the current trend of the tributaries of the Huangpu River, some artificial channels were added, and the main content of the exhibition was arranged along this circular axis. In the center of the axis are a large green area and the leisure facilities. Since the entire planning area is separated by the Huangpu River, how to transport people to the opposite bank has become one of the main considerations in this design.

The northern sector have other permanent pavilions and a huge open area. Outside the radius of the artificial lake, the south side of the Huangpu River is reserved for exhibition halls for major countries. The circle is the theme of the whole concept. Eight tower buildings will be distributed around the circle, becoming the "main pillars" that define the circular space. As part of the theme of the Shanghai World Expo 2010, the design advocates the development and utilization of reusable and recyclable materials.

The Expo will last for 184 days, and the total number of visitors is expected to reach 70 million. In order to adapt to the average daily traffic volume of about 700,000 people during peak hours, the F.A.R. of the area could reach about 4.5. The design challenge is to find a plan that can also has a long-term positive effect on the region after the Expo ends. Part of the structure will become an outdoor open space. The tower buildings outside the circular artificial river will be reserved for hotels or become residential buildings. All of the above embody the Expo theme of "Better City, Better Life".

Site area:　　　　　　　500 hm²

上海金山枫泾北美风情镇规划
Fengjing North American Style Town Planning, Jinshan, Shanghai

地点：中国，上海
业主：上海市金山区规划和自然资源管理局
设计范围：规划设计
设计时间：2003

Location: Shanghai, China
Client: Shanghai Jinshan Urban Planning and Natural Resource Administration Bureau
Design scope: Planning Design
Design time: 2003

设计在枫泾镇营造了一种生活方式。枫泾镇是江南名镇之一，已有1 500多年的历史。生态友好的多功能小镇提供了一种具有北美风情的轻松生活方式，体现了"生活方式高于建筑形式"的理念。

小镇中心区从北美特色风貌入手，80 m宽的林荫大道及门户公园组成小镇的中心区，它们与购物公园及特色果园构成北美风情生活方式的典型要素。

该方案已实施。

总用地面积:	540 hm²
总建设用地面积:	417 hm²
规划居住人口:	28 000 人

The design provided a life-style in Fengjing town. This historically important county in Jiangnan has a history of more than 1,500 years. The design of an ecologically friendly and multi-functional town offered a relaxed North American lifestyle. This philosophy is best illustrated in the words—"lifestyle is higher than architectural form".

The central core recalled the character of a typical North American town. 80 m wide boulevard and parks dominate the centre of the town. Together with the shopping park and the fruit gardens, they portrayed a classic image of North American life-style.

This design scheme has been implemented.

Site area:	540 hm²
Land used:	417 hm²
Inhabitants planned:	28,000

上海青浦中心区规划
Central Area Planning, Qingpu, Shanghai

地点：中国，上海
业主：上海市青浦城市规划管理局
设计范围：规划、实施
设计时间：2001
建成时间：2009

Location: Shanghai, China
Client: Shanghai Qingpu Urban Planning Administration Bureau
Design scope: Planning, Implementation
Design time: 2001
Completion time: 2009

沪青平高速公路出口处的水景与建筑

青浦中心区的规划始于2001年初的一个国际竞赛，KFS非常荣幸在竞赛中拔得头筹，之后即进入实施阶段。2009年，基本设计与特点初具雏形。

主要的设计元素之一是通过创建一个联结三个区域的城市绿化带来提升本区。这三个区域分别为现有老城区、新建的新城区以及展览及体育运动区域。

青浦新建的绿化带将是一个重要的生态区，除了改善空气质量的功能外，绿化带还包含一些公共空间。绿化带在新老城区间形成了一定的分隔，使人们对城市空间有了更清晰的认识。

设计的一个最大亮点是梳理现有水网而形成的夏阳湖。夏阳湖及周边已成为集青浦图书馆、博物馆和文化活动中心的城市中心区。

The planning of the Central Area of Qingpu was a competition that took place in 2001 with construction commencing shortly after the design submission of KFS was chosen as the successful entry. In 2009 the basic design and character was met.

The primary design element was the upgrading of this district through the planning and integration of green belt with those of the neighboring districts—an existing old town, a new zone and an exhibition and sport zone.

The new green belt of Qingpu served as an important ecological zone that helped to control the local air quality. It offered a number of public spaces inside itself and acted as a boundary that would give people a clearer definition of the surrounding space.

One of the biggest highlights in the design is the Xiayang Lake, which was created by modifying the existing water network. It is now a significant urban part of the central area as its banks hosted the Qingpu Library, museums and a cultural centre.

总用地面积：	452 hm^2
总建筑面积：	2 500 000 m^2
规划居住人口：	51 000 人

Site area:	452 hm^2
Gross floor area:	2,500,000 m^2
Inhabitants planned:	51,000

三亚天涯三亚湾阳光海岸规划
Sanya Bay Sunshine Coast Planning, Tianya, Sanya

地点：中国，三亚
业主：海南申亚置业有限公司
设计范围：方案设计
设计时间：2017

Location: Sanya, China
Client: Hainan Shenya Real Estate Co., Ltd.
Design scope: Proposal Design
Design time: 2017

总平面图 Master Plan

阳光海岸区块是三亚的中心城区，这一区块有极高的历史文化价值。

海南国际旅游岛逐渐成型。三亚作为海南旅游度假的热点城市，需要进一步升级城市环境。旅游城市的核心价值应随着时代变化、科技进步及文化需求的提高进行进一步升级。

总用地面积:	98 hm²
总建筑面积:	1 900 000 m²
容积率:	1.91

The Sunshine Coast is located in the downtown of Sanya. The region has an excellent historical and cultural value.

Hainan is gradually forming an international tourism island. As a popular vacation city in Hainan, Sanya needs to further improve the city environment. With the changing of times, improvements in technology and cultural demand, the core value of this tourist city should be further upgraded.

Site area:	98 hm²
Gross floor area:	1,900,000 m²
F.A.R.:	1.91

珠海横琴城市新中心规划
New City Centre Planning, Hengqin, Zhuhai

地点：中国，珠海
业主：珠海大横琴股份有限公司
设计范围：方案设计
设计时间：2017

Location: Zhuhai, China
Client: Zhuhai Dahengqin Co., Ltd.
Design scope: Proposal Design
Design time: 2017

总平面图　Site Plan

N

　　城市新中心的规划基于该区域集居住、展览、科研、商务、办公和产业为一体的功能定位，结合自然水域与绿轴，通过花园城市和街坊式布局使马骝洲水道两岸成为一体，与保税区、横琴自贸区形成产业交流。同时，设计在不同开发阶段使用空间手段进行整合，以达到提升这些区域的目的。

　　该区域为十字门商务区南湾片区先行区。该区域呈小街坊式布置，通过道路和绿化景观形成有机联系。该区域采用多元复合的发展模式，强调多元复合的城市功能，形成多种复合板块。

总用地面积：　　　　　　　4 200 hm²

The urban design of the New City Centre is based on several functions of this area—living, exhibition, scientific research, business, office and industry. It combines the natural waterscape with the greenbelt, to make both sides of Maliuzhou Waterway a garden city with streets. This design creates trade exchanges among Maliuzhou Waterway, Bonded Area and Hengqin FTZ. Meanwhile, in order to promote these places, the design integrates these areas by applying space tools in different development stages.

This area is the pioneer district of the Nanwan area of Shizimen Business District. It is arranged in a community style, organically connected through roads and landscapes. This area adopts a diversified development model, emphasizing diversified urban functions, forming multiple composite sections.

Site area:　　　　　　　4,200 hm²

上海浦东锦绣华城规划
Jinxiu City Planning, Pudong, Shanghai

地点：中国，上海
业主：上海大华集团有限公司
设计范围：规划、实施、方案设计、扩初设计
设计时间：2001
建成时间：2009

Location: Shanghai, China
Client: Shanghai Dahua Group Co., Ltd.
Design scope: Planning, Implementation, Proposal Design, Design Development
Design time: 2001
Completion time: 2009

锦绣华城是一个从规划构思到建筑实施的典型成功案例。整个设计团队用了十年时间，完成了近200万 m² 的建筑量。现今已成为浦东最大的居住区之一。

锦绣华城位于杨高路以东、博文路以南、锦绣路以西、川杨河以北，总用地面积为345 hm²，居住区用地面积为313 hm²，规划居住人口86 000人。区域内建筑种类较多，主要以居住建筑为主，还包含锦绣假日酒店、中小学校及各种类型的商业建筑等。

本项目以"泛公园"的设计理念为主，把一个345 hm²的居住区设计成一个以生态为主题的大公园，规划构筑了一个分布在345 hm²内的绿网构架。这张绿网以基地内的高压走廊、河流和城市道路为主要骨架，再把每个小区的中心绿地连接上去。

成山路是穿过该区域的主要城市道路，设有多处公交站点和地铁出入口。设计利用成山路的优势，将酒店、办公及商业服务设施集中布置在沿路两侧。商业入口广场与居住绿地的互相结合，形成独特的商业模式。

The Jinxiu City is a successful example of planning, and architectural design. Today it is one of the largest residential areas in Pudong with the completion of 2 million m² which takes the design group 10 years.

The site of the Jinxiu City is defined by four urban elements: Yanggao Road to the west, Bowen Road to the north, Jinxiu Road to the east, and Chuanyang Creek to the south. The site area is 345 hm², with 313 hm² for residential use. The planned population is 86,000. Among various types of architecture on the site, the majority is residential, combined with the elegant Holiday Inn Shanghai Jinxiu, educational facilities and different commercial units.

"Pan-Park" is the principal design element in this project. Transforming a 345 hm² residential zone into a large scale ecological-theme park. The plan is the creation of a green web within a 345 hm² plot by using high-voltage grids, rivers, and urban roads as the fundamental structure. The green zone of each neighborhood is connected to this structure.

Chengshan Road is the main traffic corridor through the site, with multiple bus stops and subway stations. To live up to this potential, the design places hotels, offices and commercial functions along this road. The main commercial entrances are merged with the green area of the residential zones, forming a model with a unique commercial character.

总用地面积：	345 hm²
总建筑面积：	2 900 000 m²
规划居住人口：	86 000 人

Site area:	345 hm²
Gross floor area:	2,900,000 m²
Inhabitants planned:	86,000

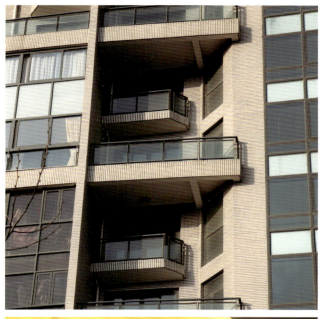

圣彼得堡波罗的海明珠规划
Baltic Pearl Planning, St. Petersburg

地点：俄罗斯，圣彼得堡
业主：波罗的海明珠股份有限公司
设计范围：规划设计
设计时间：2007

Location: St. Petersburg, Russia
Client: Baltic Pearl Co. Ltd.
Design scope: Planning Design
Design time:2007

总平面图 Site Plan

本项目整个区域的占地面积为208 hm²，本项目的设计范围是39-3地块和39a地块。

这两个地块位于整个区域的东南侧，是结合了商业和教育设施的高档居住社区。基地东侧和北侧紧邻杜杰尔戈夫斯基运河及其支流。

设计旨在延续圣彼得堡传统的城市脉络，最大化拓展周边地块价值，提升住宅价值，并通过合理的规划和设计增强社区意识。设计从每个地块的特殊性出发，以不同的设计手段对其进行分析和处理，彰显其独特的个性。设计还选择多样化的住宅种类，丰富区域的建筑语汇。

各个居住地块主要采用小尺度围合型居住组团。沿河区域主要设置联排别墅，充分利用水资源。

The entire area of the project covers an area of 208 hm². The design scope of this project is the 39–3 and 39a plots.

The two plots are located at the southeast of the site. With supporting commercial and educational facilities, they are planned to be high-end residential communities. The site borders Dugelgovsky Canal and branches to the north and east.

The design takes means to have the traditional St. Petersburg urban fabric merge into this site. Some of the goals are to maximize the value of the surrounding land in order to increase the value of the residential units and the augmentation of community-awareness through sound planning and design. Different approaches are used to analyze and process different blocks in order to give them more personality. The design chooses various types of residential units to make the community richer in architectural language and vocabulary.

Each residential plot mainly adopts small-scale enclosed residential clusters. Townhouses are mainly set up along the river to make full use of water resources.

总用地面积：	29 hm²
总建筑面积：	447 000 m²
规划居住人口：	13 000 人

Site area:	29 hm²
Gross floor area:	447,000 m²
Inhabitants planned:	13,000

马尼拉城市更新规划
Urban Regeneration Planning, Manila

地点：菲律宾，马尼拉
业主：华丽家族创新投资有限公司
设计范围：方案设计
设计时间：2018

Location: Manila, the Philippines
Client: Deluxe Family Innovation Investment Co., Ltd.
Design scope: Proposal Design
Design time: 2018

本项目位于菲律宾马尼拉大都会区的东南部城市塔吉格，东面紧邻菲律宾重要交通主干道C-5公路，北面紧邻外国使馆区，西面相距不足1 km是马尼拉都会区发展最快的博尼法西奥环球城中央商务区和城堡公园高档住宅区，距离马尼拉最大机场和马卡蒂市仅4 km。

在高密度、高容积率的条件下，设计通过大院落组合成一个大型社区空间，每户人家可以通过底层的公共走廊空间进入内部庭院或者外出。

The project is located in Taguig, a city in the southeastern part of Metro Manila, the Philippines. It is adjacent to the C–5 highway, a main road of transportation in the Philippines to the east, and a foreign embassy area to the north. The fastest growing Bonifacio Global City in Metro Manila and the upscale residential area of Castle Park are less than 1 km to the west. The site is also just 4 km away from Manila's largest airport as well as Makati City.

Under the conditions of high density and high F.A.R., large courtyards are combined into a large community space, and where each family can enter the inner courtyard or go outside through the public corridor space on the ground floor.

总用地面积：	59.0 hm²
总建筑面积：	5100000 m²
容积率：	6.41
住宅总户数：	7 217
绿化率：	40%

Site area:	59.0 hm²
Gross floor area:	5,100,000 m²
F.A.R.:	6.41
Total amount of units:	7,217
Rate of greening:	40%

哈尔滨道里爱建新城规划
Aijian New City Planning, Daoli, Harbin

地点：中国，哈尔滨
业主：哈尔滨爱达投资置业有限公司
设计范围：规划、实施、方案设计、扩初设计
设计时间：2002
建成时间：2009

Location: Harbin, China
Client: Harbin Aida Investment Real Estate Co., Ltd.
Design scope: Planning, Implementation, Planning Design, Design Development
Design time: 2002
Completion time: 2009

总平面图 Site Plan

爱建新城规划始于一个影响力很大的国际设计竞赛。KFS赢得其规划并进行了近八年的规划实施，完成了近90%的建筑设计。其中包括一系列哈尔滨标志性建筑和大量的居住建筑。

爱建新城位于松花江畔，场地原为著名的哈尔滨车辆厂。设计引入新的城市设计理念，借鉴一些可以促进经济发展、提高文化层次、提升美学和历史价值的元素，同时尊重本地的城市脉络，延续并创造富有城市情节的空间环境。设计注重保护自然生态，营造出一个绿意盎然、生机勃勃的新社区。

周围的公寓建筑围绕公园展开，围合出新城的中心地带。环形商业区外层的高层住宅面向公园，为住户提供了良好的景观。商业建筑以低层及多层为主，沿街道布置，形成亲切宜人的氛围。商业总面积约为60 hm^2。

规划路网与周边路网相协调，原规划中穿越新城的"十"字形城市主干道在新规划中予以保留。支路出口均与周边路网接通，同时地块大小尽量与周边地区的地块大小保持一致。

The planning of Aijian New City was an influential international competition. KFS placed first for the urban planning. 90% of the floor area has been completed in 8 years. This included a series of landmarks and a large amount of residential buildings.

Aijian New City is situated at the bank of Songhua River. The site's former owner was the Harbin Locomotive Factory. New concepts of urbanism are introduced using elements that elevate economic, cultural, aesthetic and historical values. The design respects the local urban fabric and establishes continuity within the urban spaces and personality. Environmental protection acts as a core principle and ensures a green and vibrant new community.

Condominiums are planned around the park, and define the centre of the community. The high-rise buildings outside the commercial ring, face the park and offer a pleasant view for the owners. The commercial part is planned with the building type limited to either low-rises or multi-storeys, evenly spread along the street to create an inviting atmosphere. 60 hm^2 is used for commercial activities.

The planned road network is in harmony with the surrounding road network, and the cross shaped urban arteries that pass through the new city in the original plan are retained in the new plan. All the branch roads are connected to the surrounding road network, and the size of the plot should be as consistent as possible with the size of the plots in the surrounding area.

总用地面积：	98 hm^2
总建筑面积：	2 200 000 m^2
建筑密度：	30%
容积率：	2.2

Site area:	98 hm^2
Gross floor area:	2,200,000 m^2
Building density:	30%
F.A.R.:	2.2

三亚吉阳亚龙湾壹号规划
1# Yalong Bay Planning, Jiyang, Sanya

地点：中国，三亚
业主：海南申亚置业有限公司
设计范围：规划、实施、方案设计、扩初设计
设计时间：2013
建成时间：2017

Location: Sanya, China
Client: Hainan Shenya Real Estate Co., Ltd.
Design scope: Planning, Implementation, Planning Design, Design Development
Design time: 2013
Completion time: 2017

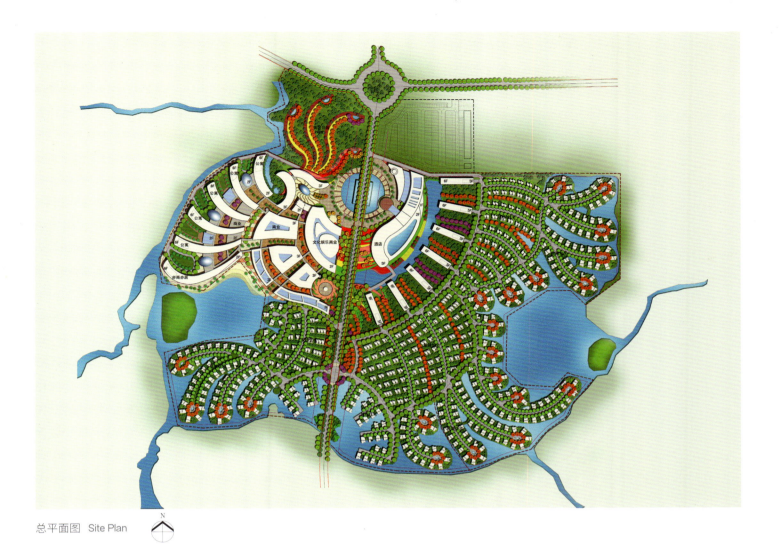

总平面图　Site Plan

本项目位于三亚亚龙湾，沿椰风路两侧布置，与周边多个五星级酒店相邻。

基地的一侧朝向大海，两边背靠青山，其中包含出现在电影《非诚勿扰》拍摄场景中的悬索桥。北侧的广场将酒店和商业两大主体连接起来，化解了椰风路两侧建筑相隔较远的不利因素。

整体的设计思路依循中国传统的"凤凰攀枝"的理念而形成。设计将五星级酒店、商业建筑和居住建筑融为一体，表现了传统的吉祥概念。不论是平视还是从附近山体鸟瞰，其独特的建筑形式必将成为当地的城市标志。

三大片区分别为五星级酒店、大型商业综合体及高档度假社区。

总用地面积:	6.76 hm²
总建筑面积:	57 000 m²
容积率:	0.84

The site is located on Yalong Bay in Sanya, spreading along Yefeng Road, and beside multiple five-star hotels.

The site is facing the sea with two sides adjacent to the mountain. There is a view of the suspension bridge on the site where the popular movie *If You Are the One* was filmed. The plaza on the north combines the hotel and the commercial zones and solves the issue of the buildings being separated on both sides of Yefeng Road.

The overall design idea is formed according to the traditional Chinese "Phoenix climbing branches" concept. The design integrates five-star hotels, commercial and residential buildings into a whole, expressing the traditional auspicious concept. The unique design which can be viewed horizontally and from a bird's view will become an iconic feature of the city.

The three areas are composed of five-stars hotels, a large commercial complex, and an upscale resort community.

Site area:	6.76 hm²
Gross floor area:	57,000 m²
F.A.R.:	0.84

西昌邛海凤凰谷度假小镇规划
Phoenix Villa Planning, Qionghai, Xichang

地点：中国，西昌
业主：成都大天实业发展有限公司
设计范围：规划设计
设计时间：2010

Location: Xichang, China
Client: Chengdu Datian Industry Co. Ltd.
Design scope: Planning Design
Design time: 2010

总平面图　Site Plan

凤凰谷度假小镇位于西昌邛海泸山景区。基地集五星级酒店、商务会议中心、休闲、居住、民俗体验与运动于一体，提供了一种舒适的生活方式。

周边的山势连绵起伏，整个用地呈扁长的"盆地形"，仅在西侧滨湖区打开一个缺口。设计结合地形，将度假区规划为三大板块：湖滨度假酒店区、风情小镇与山地高尔夫度假区。

The Phoenix Villa is situated in the Qionghai Lushan tourist zone of Xichang. The site integrated functions of a five-star hotel, a conference centre, leisure, living, local cultural experience and sports, offering a cosy life-style.

The surrounding mountains are undulating, and the entire site is in the shape of a long and flat "basin", only opening a gap in the lakeside area on the west side. In the design, combined with the topography, the resort area is planned into three sections: lakeside resort area, stylish town and mountain golf resort area.

总用地面积:	277 hm²
总建筑面积:	110 000 m²
容积率:	0.04

Site area:	277 hm²
Gross floor area:	110,000 m²
F.A.R.:	0.04

常熟虞山尚湖生态城规划
Yu Mountain Shang Lake Eco City Planning, Yushan, Changshu

地点：中国，常熟
业主：上海长甲置业有限公司
设计范围：规划设计
设计时间：2010

Location: Changshu, China
Client: Shanghai Changjia Real Estate Co., Ltd.
Design scope: Planning Design
Design time: 2010

设计在不断反省的基础上，尊重自然，注重自然生态，充分利用尚湖地区的水资源，创造出独具特色的滨水社区。在规划结构上，设计加强尚湖生态廊道和环湖路的联系，形成完整便捷的交通框架，让人们在第一时间欣赏到尚湖和虞山的主体生态美景，并自然而然地串联起各大片区。

生态之城：尚湖是常熟生活饮用水的重要水源地之一，规划保留原有湿地鸟岛等宝贵的生态资源，坚决杜绝一切对水体可能有污染的水上活动。整个规划像从尚湖自然生长出来的"水中花"。

休闲之城：结合现有度假村及水街商业，规划更加完善的滨湖度假配套设施，并以沙滩、木栈道等丰富尚湖驳岸。

文化之城：规划在滨湖沿岸增加国际会议中心、音乐厅等市政综合配套设施以及其他滨水设施。

养生之城：养生住宅、康体中心、水疗会所、高尔夫等设施使惬意的生活无处不在。

水上之城：水网发达是江南水乡的普遍优势。规划保留用地内大部分的宽阔水域，形成一座魅力四射的水上之城。

总用地面积：	300 hm²
总建筑面积：	1 300 000 m²
容积率：	0.4

The design lies on the basis of continuous reflection, respects nature, and emphasizes natural ecology. The design makes full use of the water resources in the Shang Lake area to create a unique waterfront community. As for the structure, the design strengthens the connection between Shang Lake Ecological Corridor and Huanhu Road, forming a complete and convenient transportation framework, allowing people to enjoy the main ecological beauty of Shang Lake and Yu Mountain in the first time, and naturally connect the large areas in series area.

Ecological City: Shang Lake is one of the important sources of drinking water in Changshu. The plan retains the original wetland bird island and other precious ecological resources, and resolutely eliminates all water activities that may pollute the waters. The whole plan is like a flower that naturally grows from the Shang Lake.

Leisure City: The design is the combination of holiday village with commercial functions along the lake that make for a perfect lakeside holiday experience. This experience is enhanced by beaches and wooden docks.

Cultural City: It is planned to add comprehensive municipal supporting facilities such as the International Conference Centre, Concert Hall and other waterfront facilities along the shore of the Lake.

Health City: Health-preserving residences, fitness centres, spa clubs, golf and other facilities make a comfortable life everywhere.

Water City: The concept lies in the water network, a general asset for Jiangnan. In the master plan, water retains a wide and open character to form a glamorous water city.

Site area:	300 hm²
Gross floor area:	1,300,000 m²
F.A.R.:	0.4

上海奉贤东方美谷游戏小镇规划
Dongfang Meigu Game Town Planning, Fengxian, Shanghai

地点：中国，上海
业主：上海申亚投资控股（集团）有限公司
设计范围：规划设计
设计时间：2017

Location: Shanghai, China
Client: Shanghai Shenya Investment Holding (Group) Co., Ltd
Design scope: Planning Design
Design time: 2017

近年来，中国游戏行业用户规模持续扩大，中国游戏市场在全球游戏市场中的占比稳步提升。

本项目是东方美谷游戏小镇的重要组成部分和先行启动示范区。东方美谷游戏小镇以建设传世之作为目标，充分发挥生态资源优势，借鉴荷兰羊角村的生态健康和人文特色，衔接金汇镇产业园区升级，打造集生态文明、文化创意、体验、生产、生活为一体的生态产业社区型游戏总部——"游戏羊角村"。

规划方案对原有水系进行梳理，依据现场情况，结合当地水岸特色进行水岸设计。规划区内水系与外部水系局部开通，连接水上交通。设计通过景观过渡妥善处理各栋建筑的功能关系，使区域内的各项功能衔接自然，并充分结合当地景观主要特色进行景观设计。

The gaming industry has been rapidly increasing in its number of users in China with a steadily rising percentage in the global gaming market.

The project is a vital part and a pilot zone of the Dongfang Meigu Game Town. It aims to construct a masterpiece, by fully taking advantage of the town's ecological resources, learning from the characteristics of Giethoorn in the Netherlands, linking up to the upgrade of the industrial park in Jinhui Town, building community-based gaming headquaters with eco-civilization, cultural creation, experience, manufacture and lifestyles—the "Giethoorn of Gaming".

The planning sorts out the original water system, and designs the water based on the site conditions and is combined with the characteristics of the water banks. The inner and external water systems are partially opened to connect water transportation. The functional relationship of each building is properly handled through the transitional separation of the landscape, so they are connected naturally, and the landscape design is fully integrated with the main features of the local landscape.

总用地面积：	11.99 hm²
总建筑面积：	40 550 m²
容积率：	0.33

Site area:	11.99 hm²
Gross floor area:	40,550 m²
F.A.R.:	0.33

上海徐汇滨江公共开放空间规划
Waterfront Public Space Planning, Xuhui, Shanghai

地点：中国，上海
业主：上海市徐汇区城市规划管理局
设计范围：规划设计
设计时间：2009

Location: Shanghai, China
Client: Xuhui Bureau of Urban Planning Administration, Shanghai
Design scope: Planning Design
Design time: 2009

随着城市经济和空间结构的转变，沿黄浦江两岸逐步实现产业的外移，黄浦江两岸开发进入新阶段。外滩改建、北外滩建设、南外滩建设先后实施，世博园区也已进行后续开发利用，沿江商务轴已在构建中。2011年，第14届国际泳联世界锦标赛的召开，全面点燃了黄浦江南延伸段的开发建设。

上海世博会的召开带来城市发展理念的更新和技术水平的提高。在后世博时代，"城市让生活更美好"应被注入新的诠释，并带来更加宜人的生活方式。

本项目的实施范围为日晖港至罗秀路的沿江区域。

With the transformation of urban economy and space structure, the banks of Huangpu River have entered a new phase as industries being moved out. The reconstruction of the Bund, together with the constructions in North Bund and South Bund has been conducted. The development and utilization of the World Expo area in the post-Expo era are in progress while the riverside business axis is under construction. The successful holding of the 14th FINA World Championships in 2011 has fully initiated the development of the south extended segment of Huangpu River.

The Shanghai World Expo has brought new ideas on urban development and technological improvement. New interpretation of "Better City, Better Life" should be injected into the post-Expo era to bring better lifestyles.

The scope of implementation of this project is the area along the river from Rihui Port to Luoxiu Road.

总用地面积：	24 hm²
新建建筑面积：	541 782 m²
容积率：	3.5

Site area:	24 hm²
New building area:	541,782 m²
F.A.R.:	3.5

上海长宁中山公园地区规划
Zhongshan Park Region Planning, Changning, Shanghai

地点：中国，上海
业主：上海市长宁区城市规划管理局
设计范围：规划设计
设计时间：2004

Location: Shanghai, China
Client: Changning Bureau of Urban Planning Administration, Shanghai
Design scope: Planning Design
Design time: 2004

中山公园地区位于上海地铁2号线和3号线的换乘处，是上海西部的主要商业中心。项目周边有数个公交车站并有十数条公交线路从区域中穿过，日均人流量超过40 000人。

区域周围有数个城市公园、著名大学、历史文化保护建筑、市政建筑、电影院、购物中心及高端住宅。但靠近中山公园的区域负担着大量公交线路、地铁站和一座过街天桥的交通压力。设计力图把世界上多个特大城市中心区的城市设计理念与中国本土实际相结合，打造国际级的中山公园核心区。

设计将该区域定义为四个中心。1. 交通枢纽集散中心：活力之源即大量的人流，这是中山公园地区最大的优势所在。通过立体交叉的组织方式和位于西端的环行商业步行系统，有效疏散了来自轻轨、地铁的大量瞬时人流。2. 商业购物中心：以环行广场作为区域商业中心，突破传统规划对建筑用地的限制，空间自由度得到释放，商业活动潜能得到极致挖掘。3. 信息中心：统筹布置的广告、招贴、灯光工程，形成流光溢彩、目不暇接的大容量信息港湾。4. 商务酒店、商业活动中心：创造更加适宜商业办公的写字楼群，并与保留建筑交相辉映，其风格定位与虹桥商务区差别化，互为补充。

总用地面积：	14.5 hm²
总建筑面积：	510 700 m²

Located at the interchange of Metro Line 2 and Line 3 of Shanghai, the Zhongshan Park region is the main commercial centre of western Shanghai. There are several bus stations around the project and dozens of bus lines pass through the area, with an average daily flow of more than 40,000 people.

The site features a variety of city parks, famous universities, buildings of cultural heritage, government buildings, cinemas, shopping malls and prestigious residential areas. However, the area near Zhongshan Park is burdened with heavy bus traffic, metro stations and an overpass. The design strives to combine the urban design concepts of many megacities in the world with the local reality of China to create a world-class Zhongshan Park core area.

The urban renewal design defines the region as four centres. 1.Transportation distribution centre: The source of vitality is people, the biggest advantage of Zhongshan Park region. Through the three-dimensional intersection and the circular commercial pedestrian system located at the west end, a large amount of instantaneous pedestrian flow from light rail and subway is effectively evacuated. 2. Commercial shopping centre: the circular plaza is used as the regional commercial centre, to break through the limitations of traditional planning so that the freedom of space is released and the potential of commercial activities is fully explored. 3. Information centre: Coordinately arranged advertisements, posters, and lighting projects, forming a glamorous and dizzying large-capacity information harbor. 4. Business hotels and commercial activity centre: create office buildings that are more suitable for commercial offices, whose style positioning is differentiated from that of the Hongqiao Business District and complement each other.

Site area:	14.5 hm²
Gross floor area:	510,700 m²

PUBLIC DESIGN

公共建筑

珠海横琴东西汇综合体
Complex of WE Park, Hengqin, Zhuhai

地点：中国，珠海
业主：东西汇（横琴）发展有限公司
设计范围：规划设计、方案设计、扩初设计
设计时间：2019

Location: Zhuhai, China
Client: WE Park (Hengqin) Development Co., Ltd.
Design scope: Planning Design, Proposal Design, Design Development
Design time: 2019

总平面图 Site Plan

本项目的整体概念来源于东西方文化的交融，整体形态上体现了东方山水画和西方舞蹈的意向。

方案由沉浸式1 500座剧场、红唇晚宴剧场、音乐餐厅和实景剧场等文化旅游设施，以及文化创意产业、办公、商业等功能构成。项目内设有娱乐餐饮，打造多功能、现代化的商业综合体，为消费者带来全新的消费概念和生活体验，打造文化旅游目的地以及文化创意产业平台。

南北商业步行街以"龙"为意向设计，屋顶犹如鳞片，晚上与塔楼形成变幻多彩的灯光空间。商业整体以绿化园林为主题，配合自然景观、绿墙和大量自然光线，营造出购物的气氛。设计引入绿化广场作为漫步的场地和购物中心的核心点，并与商业街相结合。设计除了提供多层次的活动空间，也把戏剧以及东西文化引入各商铺。

The overall concept of this project originates from the integration of Eastern and Western cultures. The overall form represents the eastern painting and the western dance.

The scheme consists of cultural tourism facilities such as immersive theatre with 1,500 seats, red lips dinner theatre, music restaurant and live theatres, as well as cultural and creative industries, offices and commercials. The project contains entertainment and catering to create a modern multi-functional commercial complex. It provides consumers with a brand-new concept and life experience, to create a cultural tourism destination and a cultural industry platform.

The commercial pedestrian street is designed with the imagery of dragon. The canopies form a colourful lighting space with the tower at night. The commercial space is based on the theme of gardens, combined with natural landscapes, green walls and natural light to create an atmosphere for shopping. It introduces green squares for walking as the core of shopping centres and integrates with commercial streets. In addition to providing multi-level activity spaces, the design also introduces opera, eastern and western cultures to various shops.

总用地面积：	8.4 hm²
总建筑面积：	373 000 m²
容积率：	2.6
文化创意产业面积：	76 000 m²
办公面积：	84 000 m²
商业面积：	50 000 m²

Site area:	8.4 hm²
Gross floor area:	373,000 m²
F.A.R.:	2.6
Cultural and creative industry area:	76,000 m²
Office area:	84,000 m²
Commercial area:	50,000 m²

一层平面图 1st Floor Plan

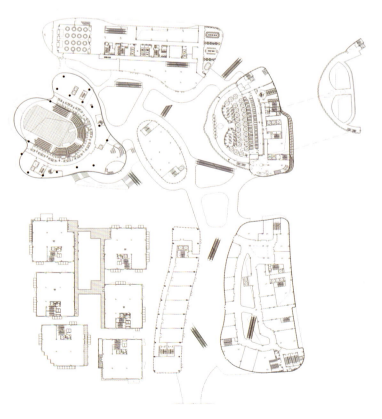

二层平面图 2nd Floor Plan

三层平面图 3rd Floor Plan

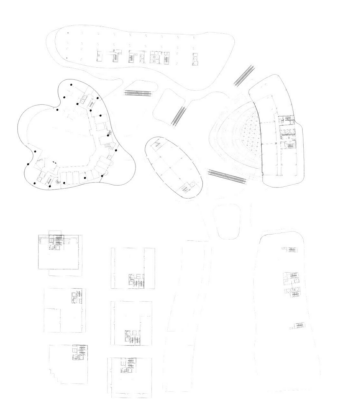

四层平面图 4th Floor Plan

珠海横琴"天沐琴台"综合体
Complex of "Tianmu Qintai", Hengqin, Zhuhai

地点：中国，珠海
业主：珠海市横琴新区管理委员会
设计范围：策划研究、规划设计、方案设计
设计时间：2010

Location: Zhuhai, China
Client: Zhuhai Hengqin New Area Management Council
Design scope: Strategic Research, Planning Design, Proposal Design
Design time: 2010

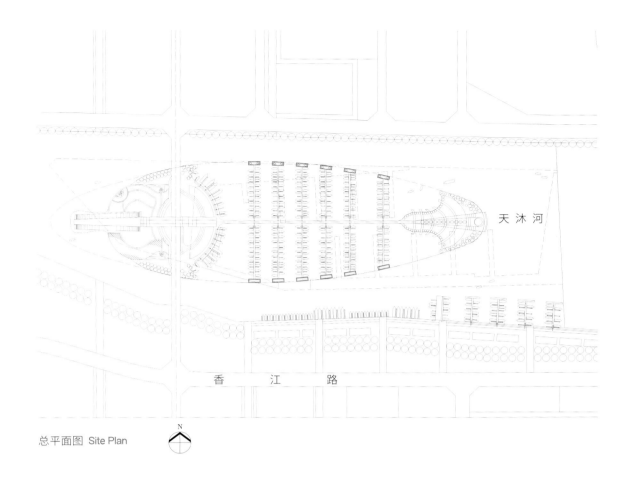

天沐河

总平面图 Site Plan

N

香 江 路

"天沐琴台"综合体位于珠海横琴，其地理位置景观环境优越。

设计立意为横琴——情岛。横琴毗邻香港和澳门，这里原来是两个岛：大横琴岛和小横琴岛。因其地形和山势，大小二岛像横在南海碧波上的两具古琴，千百年来，日日夜夜和着山风与海涛弹奏着山之歌、风之歌与海之歌。

设计结合其地理位置特色，以横置的中国古琴的形态作为建筑的基本形式。

The site for the Complex of "Tianmu Qintai" is located at Hengqin, Zhuhai. The geographical feature provides the site with stunning landscapes.

The design idea is inspired by Hengqin, literally meaning horizontal lute—Island of romance; being adjacent to Hong Kong and Macao, it was originally two islands: Big Hengqin Island and Small Hengqin Island. These two parallel islands resemble a couple of Chinese ancient lutes, eternally playing music for the mountain, the wind, and the sea throughout the day.

The design combines all these geographical features to result in a form of a horizontally positioned Chinese ancient lute.

总用地面积：	3.5 hm²
地上建筑面积：	68 000 m²
容积率：	1.90

Site area:	3.5 hm²
Above ground G.F.A.:	68,000 m²
F.A.R.:	1.90

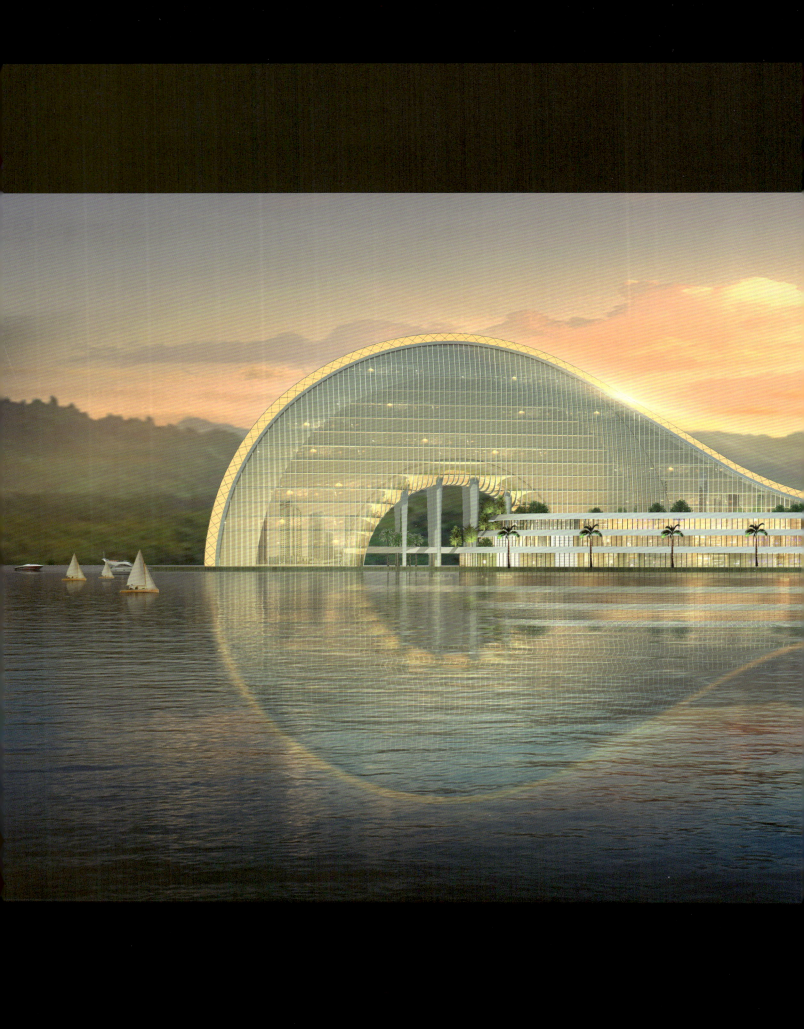

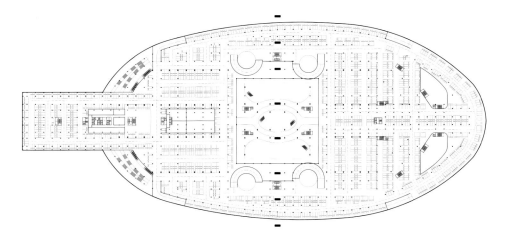

地下二层平面图 B2 Floor Plan

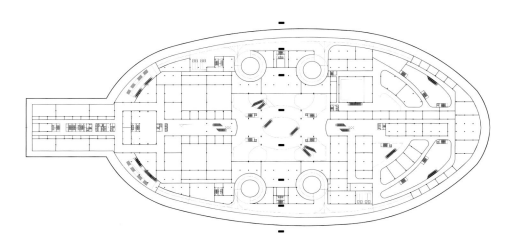

地下一层平面图 B1 Floor Plan

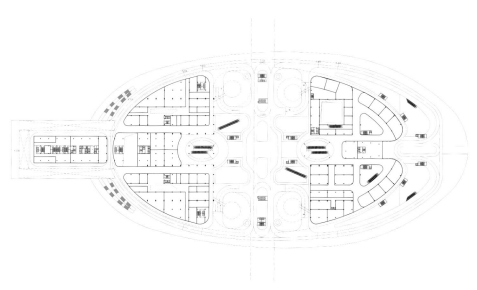

一层平面图 1st Floor Plan

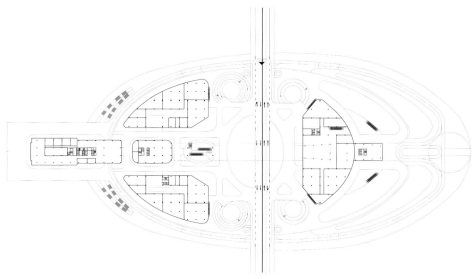

二层平面图 2nd Floor Plan

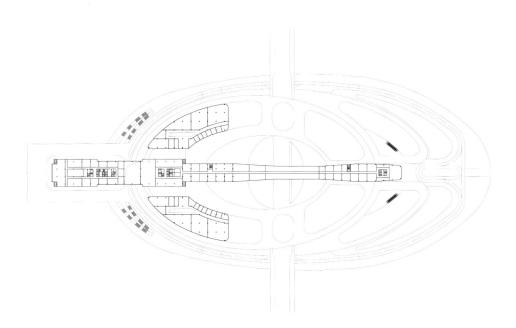

三层平面图 3rd Floor Plan

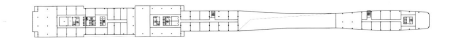

四层平面图 4th Floor Plan

珠海香洲珠海市民服务中心
Zhuhai Civic Service Centre, Xiangzhou, Zhuhai

地点：中国，珠海
业主：珠海市人民政府
设计范围：方案设计
设计时间：2020

Location: Zhuhai, China
Client: Zhuhai Municipal People's Government
Design scope: Proposal Design
Design time: 2020

凤凰山

待建隧道口

梅 华 西 路

人行入口

工作人员入口 工作人员入口 工作人员入口

13F 12F

13F 12F

工作人员入口

5F

13F

H

公交车首末站

应急中心入口

应急中心停车场

彩 虹 路

大厅入口

会议区入口

青 活 路

车行入口

车行入口

车行入口

迎 宾 北 路

总平面图 Site Plan N

　　本项目位于珠海新香洲片区，用地紧邻城市的行政和体育中心，毗邻凤凰山和大海，自然资源极其丰富。

　　本项目周边环境良好，配套设施完善，交通可达性较高。本项目在城市风貌形象的展示与市民服务功能的提升等方面将发挥重要作用。

　　设计首先确立了规划和景观的设计思路，初步建立起市民中心，如"溅起的水花"般的建筑意境。场地分为市民中心、应急中心和公交车首末站三大主要分区。其中，市民中心包括办事大厅、行政审批办公室、会议配套设施及便民服务等主要功能。

This project is located in the New Xiangzhou Area of Zhuhai. The site is close to the city administration and sports centre. It is adjacent to Phoenix Mountain and the sea with abundant natural resources.

The surrounding area of the project is built with a pleasant environment and a complete infrastructure. The traffic is highly accessible. The project will play an important role in the display of the city image and the improvement of the citizens' service function.

The design first established planning and landscape design ideas and built up the architectual conception of "splashing water" of the Civic Centre. The site is divided into three main zones: the Civic Centre, the Emergency Centre, and the bus terminal. The Civic Centre includes the main functions of the service hall, the administrative approval offices, the conference facilities and the convenient services for the public.

总用地面积:	4.36 hm²
总建筑面积:	81 000 m²
容积率:	1.86

Site area:	4.36 hm²
Gross floor area:	81,000 m²
F.A.R.:	1.86

珠海香洲珠海市文化艺术中心
Zhuhai Cultural & Art Centre, Xiangzhou, Zhuhai

地点：中国，珠海
业主：珠海市人民政府
设计范围：方案设计
设计时间：2020

Location: Zhuhai, China
Client: Zhuhai Municipal People's Government
Design scope: Proposal Design
Design time: 2020

总平面图 Site Plan

设计采用流线型、非线性的浪花造型突出新时代的建筑特征，通过建筑本身体现时代特征和改革开放的气质和形象，体现出一个带有强烈地缘特征的亲民和谐的珠海城市人文地标。

建筑主体沿海岸展开。根据项目特征，建筑被划分为四个体量——群众文化演艺中心、主题展览馆、培训创作区和市图书馆。它们面朝大海，创造出各种类型的共享空间，并且都拥有非常好的沿海景观。用地西北部靠近城市主干道，适合布置公共的集散广场，并为市民提供聚集活动的场所；用地东部紧贴半岛沙滩，提供了独特的海岸线景观。

The design adopts streamlined, non-linear gushing waves to highlight the characteristics of the architecture of the new era. Through the building itself, it reflects the characteristics of the times, the temperament and the image of the reform and opening-up. Also the project is a harmonious and friendly cultural landmark of Zhuhai with strong geographical characteristics.

The main body of the building unfolds along the coast. According to the characteristics of the project, the building is divided into four volumes—the Public Performing Centre, the Theme Exhibition Hall, the Training & Creation Area and the City Library, all facing the sea to create various types of shared spaces, each of which has a very good coastal landscape. The northwest of the site is close to the main road, which is suitable for arranging a public distribution square, providing the citizens with a place for gathering activities. The east of the site is close to the peninsula beach, providing a unique landscape of the coastline.

总用地面积:	28.78 hm²
总建筑面积:	199 859 m²
容积率:	0.69

Site area:	28.78 hm²
Gross floor area:	199,859 m²
F.A.R.:	0.69

上海嘉定文博苑
Cultural Museum Complex, Jiading, Shanghai

地点：中国，上海
业主：上海市嘉定区规划和土地管理局
设计范围：方案设计
设计时间：2015

Location: Shanghai, China
Client: Shanghai Jiading Urban Planning and Land Administration Bureau
Design scope: Proposal Design
Design time: 2015

上海嘉定文博苑
Cultural Museum Complex, Jiading, Shanghai

地点：中国，上海
业主：上海市嘉定区规划和土地管理局
设计范围：方案设计
设计时间：2015

总平面图 Site Plan

本项目位于上海嘉定老城区南端，集文化、旅游、休闲和园林四方面为一体，主题功能包括博物馆和美术馆。

设计构思来源于从天而降的宝石，其撞击地面释放出无限能量，并成为"嘉定之门"。参观者可以从中央核心处的主入口进入三座建筑，在这里聚集并经垂直路径到不同的建筑参观。次入口为博物馆工作人员使用的内部人流和货运后勤入口。参观者由西侧的入口大厅步入中庭，并在这里参观围绕中庭安排的多个展览空间。藏品通过专用货梯从地下室进入各展厅。

The project is located in the southern part of Jiading old town, Shanghai. The project contains culture, tourism, leisure and garden four aspects with museum and art gallery as the main functions.

The design concept is inspired by the falling gemstone from the sky which reveals its energy when hits the ground and becomes "the gate of Jiading". The visitors could enter all three buildings from the main entrance in the central core where the visitors gather together and travel vertically to different buildings. The secondary entrance is for museum staff use of internal flow and logistics transportation. The visitors enter the main hall from the west and explore the different exhibition spaces which are arranged around the atrium. The collection is transferred by private ladder from the basement to the exhibition halls.

总用地面积：	3.45 hm²
地上建筑面积：	81 864.7 m²
容积率：	2.37

Site area:	3.45 hm²
Above ground G.F.A.:	81,864.7 m²
F.A.R.:	2.37

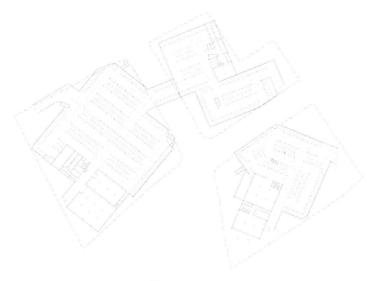

一层平面图 1st Floor Plan

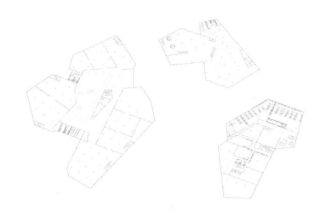

二层平面图 2nd Floor Plan

三层平面图 3rd Floor Plan

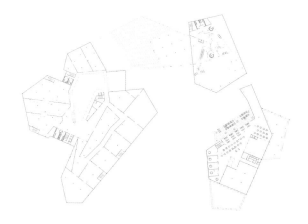

四层平面图 4th Floor Plan

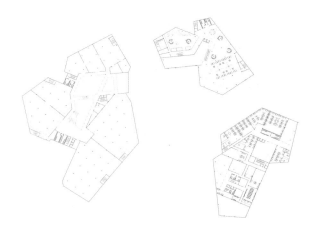

五层平面图 5th Floor Plan

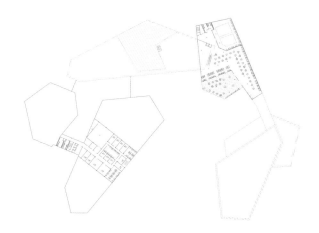

六层平面图 6th Floor Plan

珠海横琴十字门中央商务区教育设施
Shizimen CBD Educational Facilities, Hengqin, Zhuhai

地点：中国，珠海
业主：珠海十字门中央商务区建设控股有限公司
设计范围：方案设计
设计时间：2020

Location: Zhuhai, China
Client: Zhuhai Shizimen CBD Construction Holdings Co., Ltd.
Design scope: Proposal Design
Design time: 2020

本项目位于珠海横琴十字门中央商务区。基地南侧为马骝洲水道，地理位置极佳。

本项目为包含一所18班幼儿园、一所36班小学、一所48班初中及一所36班高中的教育设施方案设计。项目定位为城市教育地标，有着得天独厚的地段和景观优势。

设计采用"步步高升"的总体概念，将四项主要功能沿中轴线布置。设计将主要建筑体量布置在基地西侧，减少山体对视线的干扰。教室全部被布置在建筑的南侧，保证每间教室都拥有良好的视线和日照，并满足采光和通风要求。设计在每栋教学楼楼顶都布置了室外活动平台和观景点，营造了垂直景观，丰富了空间层次，形成了丰富的空间体验。

The project is located in Shizimen CBD of Hengqin, Zhuhai. The south side of the base is the Maliuzhou Waterway. The geographical position is excellent.

This project is a conceptual layout design of educational facilities for a 18-class kindergarten, a 36-class elementary school, a 48-class junior high and a 36-class senior high. The project is positioned as an urban educational landmark, with unique location and landscape advantages.

The design generally adopts the concept of "rising step by step". The four main functions are distributed along the central axis. The design arranges main building volumes on the west side of the site to reduce the interference of the mountain. The classrooms are all arranged on the south side of the building to ensure good views and sunshine of each classroom, and to meet the lighting and ventilation regulations. Outdoor platforms and viewing points are arranged on top of each building, to create vertical landscapes, enrich the spatial layers and form a complete spatial experience.

总用地面积：	11.08 hm²
总建筑面积：	168 581 m²
容积率：	1.52

Site area:	11.08 hm²
Gross floor area:	168,581 m²
F.A.R.:	1.52

珠海横琴南山咀地块
Nanshanzui Lot, Hengqin, Zhuhai

地点：中国，珠海
业主：珠海市横琴新区管理委员会规划国土局
设计范围：方案设计
设计时间：2020

Location: Zhuhai, China
Client: Bureau of Planning, Land and Resources of Zhuhai Hengqin New Area Manangement Council
Design scope: Proposal Design
Design time: 2020

上海浦东大华锦绣假日酒店
Holiday Inn Dahua Jinxiu, Pudong, Shanghai

地点：中国，上海
业主：上海大华集团有限公司
设计范围：方案设计、扩初设计
设计时间：2005
建成时间：2010

Location: Shanghai, China
Client: Shanghai Dahua Group Co., Ltd.
Design scope: Proposal Design, Design Development
Design time: 2005
Completion time: 2010

总平面图 Site Plan

N

大华锦绣假日酒店位于上海浦东，属于锦绣华城社区。酒店西面紧邻锦尊路，南面与北面皆为商业建筑设施用地，东面为住宅区。整个酒店以体育公园为主要朝向，使最多的房间能面向最佳的景观。南面的商业设施与酒店互相烘托人气，方便客人使用。

21层的酒店地块呈细长的L形。酒店地下层主要为酒店车库和管理用房；4层至21层为酒店客房区；1层至3层主要设有24小时餐厅、特色餐厅、500人宴会厅、会议室、游泳池、水疗等休闲娱乐设施。便捷的交通与商务人士快节奏的生活十分合拍。各种高端的设施满足了客户休闲和娱乐的要求。

The Holiday Inn Dahua Jinxiu is located in Pudong, Shanghai, and belongs to Graceful Oasis City. The hotel stands east of Jinzun Road, north and south of the commercial facilities and west of the residential area. The whole hotel is mainly oriented towards the sports park, so that the most rooms can face the best view. The commercial facilities in the south and the hotel complement each other's popularity and are convenient for the guests to use.

The site for the 21-storey hotel is an elongated L-shape. The parking area and the management offices of the hotel are located underground. The hotel rooms occupy the 4th floor to the 21st floor. The hotel provides a 24-hour restaurant, a theme restaurant, a 500-seat banquet, conference rooms, swimming pool, spa and other recreational facilities from the 1st floor to the 3rd floor. The convenient transportation suits the fast pace of businessman. A variety of high-end facilities meet the customer's leisure and entertainment requirements.

总用地面积：	1.1 hm²
总建筑面积：	41 000 m²
地上建筑面积：	32 000 m²
容积率：	2.9
酒店客房数：	219

Site area:	1.1 hm²
Gross floor area:	41,000 m²
Above ground G.F.A.:	32,000 m²
F.A.R.:	2.9
Hotel rooms:	219

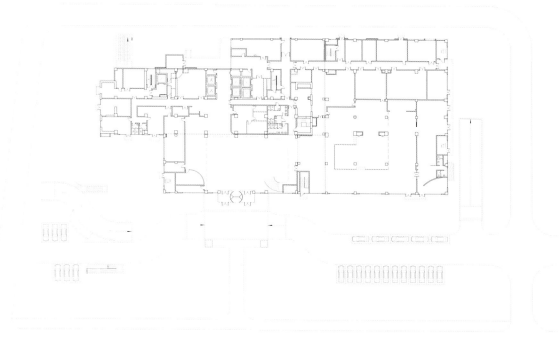

一层平面图 1st Floor Plan

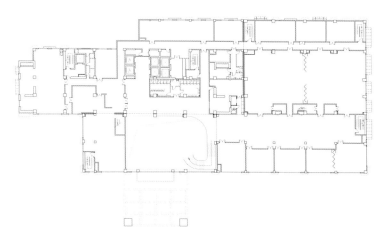

二层平面图 2nd Floor Plan

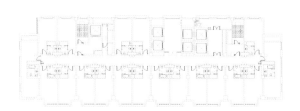

标准层平面图 Typical Floor Plan

三亚吉阳亚龙湾AC-4地块美高梅酒店
Yalong Bay Lot AC–4, MGM Hotel, Jiyang, Sanya

地点：中国，三亚
业主：海南申亚置业有限公司
设计范围：方案设计、扩初设计
设计时间：2017

Location: Sanya, China
Client: Hainan Shenya Real Estate Co., Ltd.
Design scope: Proposal Design, Design Development
Design time: 2017

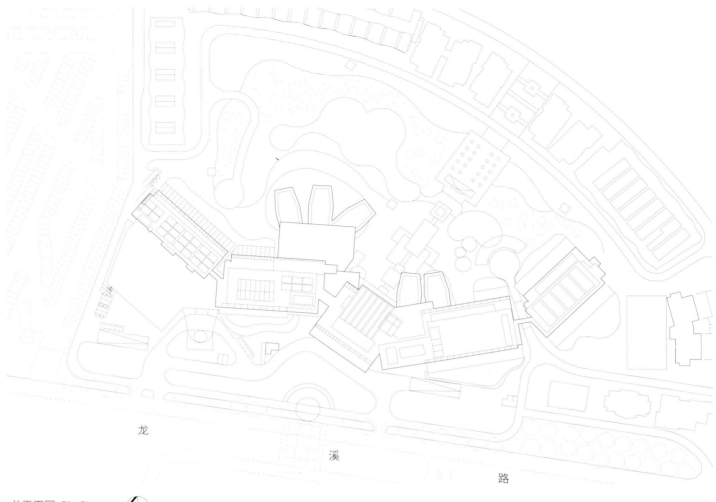

总平面图 Site Plan　N

本项目位于三亚二线海岸。设计在总体规划时，考虑了酒店建筑与商业综合体的联系以及其他建筑对酒店的视线影响，同时强调了中心景观区的打造。设计在室外区域打造了丰富的景观和泳池资源，力争将室外环境变成酒店的一大亮点。

项目定位为富有活力、娱乐性和吸引力的热带旅游度假酒店，将美高梅著名的娱乐体验带入本项目当中。

设计延续并呼应了三期地块的轴线，将酒店的大堂与入口设置在该轴线上。受建筑高度的限制，建筑设计为五层，主体建筑沿南北方向呈板式布局展开，形成完整的沿街界面。主要客房沿东西向布置，与三期建筑保持合理间距。酒店与三期公寓之间形成大面积景观内院空间。

The project is located on the second-tier coast of Sanya. In the master plan, the connection between the hotel buildings and the commercial complex, and the sight interference of other buildings are considered. The construction of the central landscape is emphasized, creating rich landscape and pool resources, striving to make the outdoor environment a highlight.

The project is positioned as a dynamic, entertaining and attractive tropical tourism resort, bringing MGM's famous entertainment experience into the project.

The design continues and echoes the axis of the third phase, and sets the hotel lobby and entrance on this axis. Limited by the height of the building, the building is designed to be five storeys. The main building unfolds in the north-south direction in a plate layout, forming a complete interface along the street. The main hotel rooms are arranged in the east-west direction, keeping a reasonable distance with the third phase buildings. A large area of landscape courtyard space is formed between the hotel and the third phase apartments.

总用地面积：	4.5 hm²
总建筑面积：	33300 m²
容积率：	0.74
酒店客房数：	353

Site area:	4.5 hm²
Gross floor area:	33,300 m²
F.A.R.:	0.74
Hotel rooms:	353

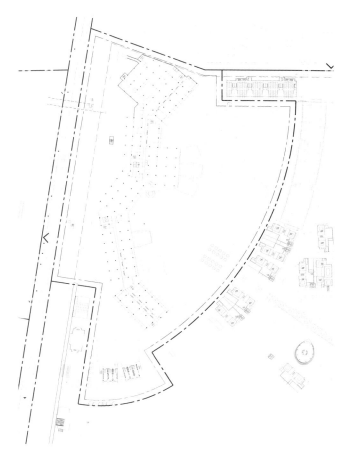

地下一层平面图 B1 Floor Plan

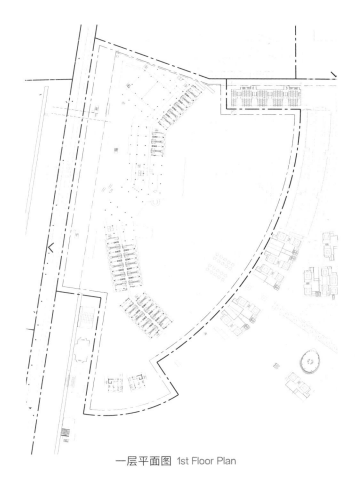

一层平面图 1st Floor Plan

龙溪路立面图 Elevation along Longxi Road

东侧内院立面图 Elevation along East Yard

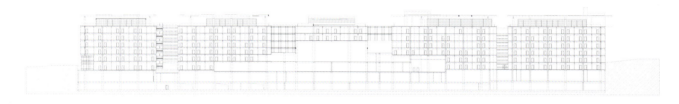

1—1剖面图 1—1 Section

2—2剖面图 2—2 Section

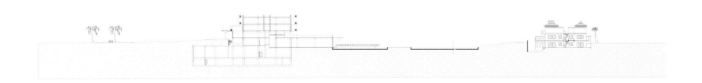

3—3剖面图 3—3 Section

无锡新吴无锡千禧大酒店
Wuxi Millennium Hotel, Xinwu, Wuxi

地点：中国，无锡
业主：无锡鑫畅置业有限公司
设计范围：方案设计、扩初设计
设计时间：2006
建成时间：2009

Location: Wuxi, China
Client: Wuxi Xinchang Real Estate Development Co., Ltd.
Design scope: Proposal Design, Design Development
Design time: 2006
Completion time: 2009

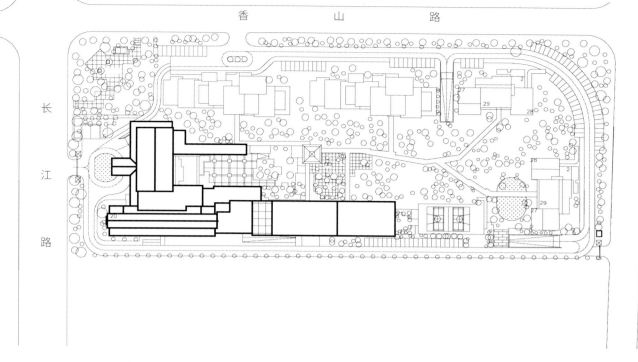

总平面图 Site Plan

本项目位于无锡市新吴区，西侧是长江路，北侧是香山路。酒店共两栋建筑，总建筑面积为35 000 m²。

酒店主出入口设置在长江路上，主出入口南侧设置机动车出入口，两个出入口与公寓区的主出入口互不干扰。地下室为机动车库、设备用房、娱乐休闲及服务用房；一、二层主要包括健身房、餐厅及会议室等；三层至二十二层为酒店客房。

酒店建筑拥有一个约40 m宽、150 m长的景观花园。半封闭式的"景观回廊"连接了花园以及基地内的各栋建筑，使酒店能更好地为不同人群服务，并提高了酒店的利用率。同时，"景观回廊"形成各主题景观空间，体现了中式园林的特色。

在设计中，KFS坚持建筑、室内的一体化设计，完成了酒店所有的建筑设计和室内设计。

总用地面积：	1.1 hm²
总建筑面积：	35 000 m²
容积率：	2.9
酒店客房数：	306

The project is located in the Xinwu District of Wuxi, and it is on the east of Changjiang Road and south of Xiangshan Road. The site supports two hotel buildings with a combined gross floor area of 35,000 m².

The main entrance of the hotel is located at Changjiang Road with the vehicle entrance to the south. The main entrance of the apartment blocks is set apart from the former two entrances. The underground level of the hotel provides underground parking places, equipment rooms, recreational rooms and service rooms. The gyms, the restaurants and the conference rooms are located on the first and second floors. The hotel rooms are on the third to the twenty-second floors.

The hotel buildings provide a landscape garden about 40 m wide and 150 m long. The garden and other buildings in the base can be reached through a semi-enclosed "landscape corridor", so that the hotel can better serve different people and improve its utilization rate. At the same time, various themed landscape spaces are formed through the "landscape corridor", which embodies the characteristics of Chinese gardens.

In the design, KFS insisted on the integrated design of architecture and interior, and completed all the architectural design and interior design of the hotel.

Site area:	1.1 hm²
Gross floor area:	35,000 m²
F.A.R.:	2.9
Hotel rooms:	306

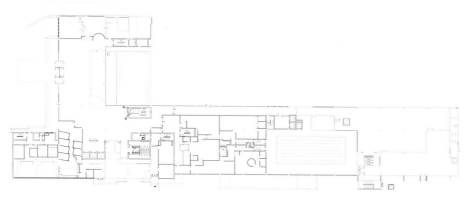

一层平面图 1st Floor Plan

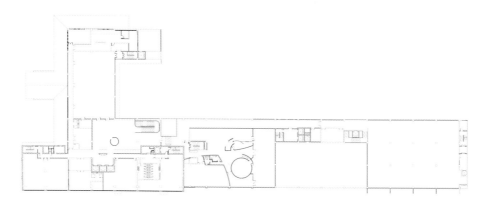

二层平面图 2nd Floor Plan

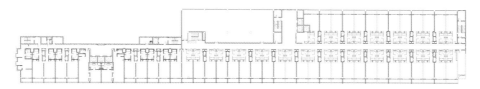

三至四层平面图 3rd to 4th Floor Plan

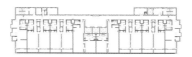

标准层平面图 Typical Floor Plan

成都武侯新东方千禧大酒店
New Orient Millennium Hotel, Wuhou, Chengdu

地点：中国，成都
业主：成都新东方置业有限公司
设计范围：方案设计、扩初设计
设计时间：2004
建成时间：2009

Location: Chengdu, China
Client: Chengdu New Orient Real Estate Inc.
Design scope: Proposal Design, Design Development
Design time: 2004
Completion time: 2009

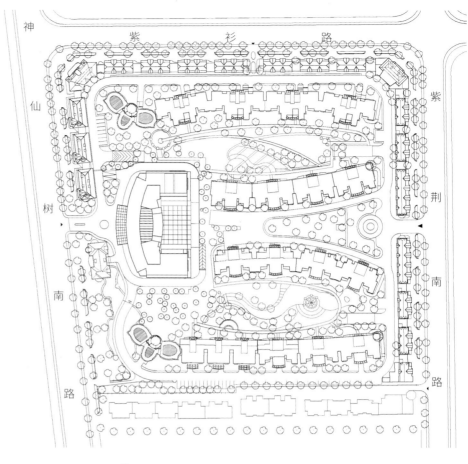

总平面图 Site Plan

本项目位于成都高新技术产业开发区内，为一家高标准的五星级酒店。

酒店主楼共14层，裙房2层，地下1层，另有商务会议、餐饮、休闲等设施，是成都高标准、多功能的五星级酒店。

基地位于上海花园小区内，并紧邻城市道路。酒店西南处有一个面积约2 000 m²的市政公园，为酒店提供了良好的景观。酒店的设计考虑到最大限度地利用景观资源，有80%的房间可以看到绿地。酒店内部设计有一个底层至11层的大型中庭共享空间，它配备有独立的车行和人行系统以避免来自周边居住区的干扰。新古典主义的建筑风格采用了简洁的处理以使之与周边的现代建筑相协调。

The project is a high standard five-star hotel located in the High-Tech Industrial Development Zone of Chengdu.

The building is fourteen storeys, with two-storey podiums and one-storey underground. It is equipped with conference rooms, restaurants and entertainment facilities. The building is designed to be a high quality, multi-use five-star hotel in Chengdu.

The site is situated inside Shanghai Garden and enclosed by streets. The 2,000 m² city park is in southwest of the hotel, supplying great views to the hotel. The design considers the maximum use of lanscape resources, making 80% of the rooms face the open greenery. The interior of the hotel is designed with a large atrium shared space from the ground floor to the 11th floor, which is equipped with independent vehicle and pedestrian systems to avoid interference from the surrounding residential areas. The neoclassical architectural style adopts a simple treatment to make it coordinate with the surrounding modern buildings.

总用地面积：	0.8 hm²
总建筑面积：	38 000 m²
容积率：	3.7
酒店客房数：	319

Site area:	0.8 hm²
Gross floor area:	38,000 m²
F.A.R.:	3.7
Hotel rooms:	319

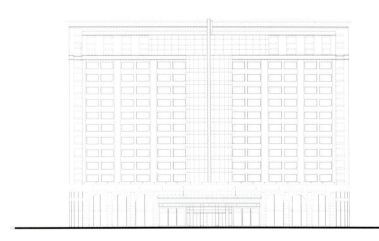

立面图 Elevation

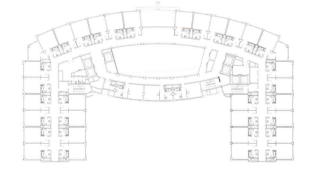

标准层平面图 Typical Floor Plan

珠海横琴中大金融大厦
Zhongda Financial Tower, Hengqin, Zhuhai

地点：中国，珠海
业主：中大控股有限公司
设计范围：方案设计、扩初设计
设计时间：2014
建成时间：2019

Location: Zhuhai, China
Client: Zhongda Holdings Co., Ltd.
Design scope: Proposal Design, Design Development
Design time: 2014
Completion time: 2019

本项目位于珠海横琴十字门中央商务区十字门大道东侧、汇通五路南侧、荣港道西侧、汇通三路北侧。用地靠近离岸金融岛中心绿地，东北方向可远眺澳门。

本项目与澳门隔海相望，海景条件优越。项目由一栋250 m高的商务办公楼和两栋100 m高的商务公寓楼及商业裙房所组成。

The project is located at Shizimen CBD of Hengqin, Zhuhai. It is surrounded by Shizimen Avenue on the west, Huitongwu Road on the north, Ronggang Avenue on the east, and Huitongsan Road on the south. It is close to the central green land of the Off-shore Financial Centre, and the far northeast is Macao.

The geographic location with Macao across the sea provides an outstanding ocean view. The project combines a 250 m high commercial office tower, two 100 m high commercial apartments and commercial podiums.

总用地面积：	1.87 hm²
总建筑面积：	111 000 m²
容积率：	5.80

Site area:	1.87 hm²
Gross floor area:	111,000 m²
F.A.R.:	5.80

裙楼一层平面图 Podium 1st Floor Plan

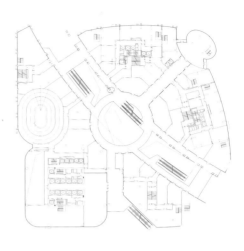

裙楼二层平面图 Podium 2nd Floor Plan

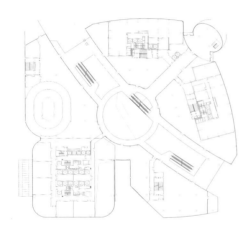

裙楼三层平面图 Podium 3rd Floor Plan

裙楼四层平面图 Podium 4th Floor Plan

裙楼五层平面图 Podium 5th Floor Plan

裙楼六层平面图 Podium 6th Floor Plan

地下一层平面图 B1 Floor Plan

地下二层平面图 B2 Floor Plan

地下三层平面图 B3 Floor Plan

塔楼典型平面图 Typical Floor Plan of the Tower

上海浦东张江长泰国际广场
Zhangjiang Changtai International Plaza, Pudong, Shanghai

地点：中国，上海
业主：上海金缔联创置业有限公司
设计范围：方案设计、扩初设计
设计时间：2010
建成时间：2014

Location: Shanghai, China
Client: Shanghai Jindi Lianchuang Real Estate Co., Ltd.
Design scope: Proposal Design, Design Development
Design time: 2010
Completion time: 2014

总平面图 Site Plan

本项目位于上海浦东。总用地面积为8.2 hm²，北部为高层办公楼，南部为商业广场。

商业广场的外观设计以整齐简洁的新古典风格为主，局部配以现代风格立面体量及元素，体现传统与现代的碰撞与融合。在这里，古典的典雅高贵与现代的激情时尚同时呈现，符合现代商业的发展方向。

This project is located in Pudong, Shanghai. The site area is 8.2 hm², with high-rise office buildings in the north and a commercial plaza in the south.

The exterior design of the commercial plaza is dominated by a neat and concise neoclassical style, partially equipped with modern-style facade volumes and elements, reflecting the collision and fusion of tradition and modernity. Here, the classical elegance and nobility are presented at the same time as the modern passion and fashion, which is in line with the development direction of modern business.

总用地面积：	8.2 hm²
总建筑面积：	165 000 m²
容积率：	2.00

Site area:	8.2 hm²
Total floor area:	165,000 m²
F.A.R.:	2.00

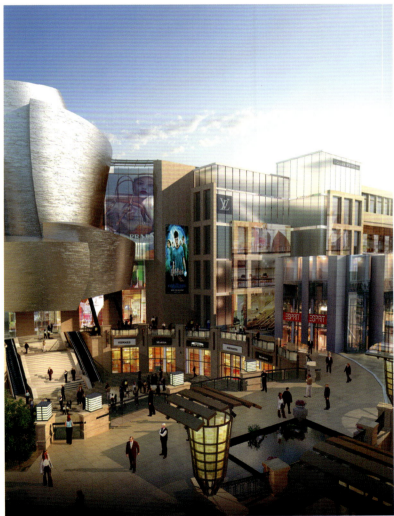

三亚吉阳亚龙湾壹号商业综合体
Commercial Complex of 1# Yalong Bay, Jiyang, Sanya

地点：中国，三亚
业主：海南申亚置业有限公司
设计范围：方案设计、扩初设计
设计时间：2013
建成时间：2016

Location: Sanya, China
Client: Hainan Shenya Real Estate Co., Ltd.
Design scope: Proposal Design, Design Development
Design time: 2013
Completion time: 2016

总平面图 Site Plan

水　面

龙

溪

路

本项目与周边的多个五星级酒店相邻。本地块位于进入三亚亚龙湾的主干道椰风路的两侧。

两大城市门户分别位于项目南北两端，极具视觉冲击力。其中，北侧的广场将酒店和商业两大主体连接成一个整体，化解了椰风路两侧建筑相隔较远的不利因素。

整体商业建筑在考虑安全舒适及方便购物的基础上适当地引入了当地的传统戏剧文化元素，背篓形状的演绎展现了独特的建筑形式，成为当地的城市商业标志。

The project is surrounded by multiple five-star hotels. The site spreads along Yefeng Road, which is the main entrance corridor of Yalong Bay.

Two urban gateways are located at the north and south ends of the project, which have a great visual impact. Among them, the plaza on the north side connects the two main bodies of the hotel and the commercial into a whole, which resolves the disadvantage of the distance between the buildings on both sides of Yefeng Road.

The overall commercial building introduces local traditional drama and cultural elements on the basis of safety, comfort and convenience for shopping. The interpretation of the basket shape shows a unique architectural form, which has become a local commercial landmark of the city.

总用地面积:	6.76 hm²
总建筑面积:	93 506 m²
容积率:	0.84

Site area:	6.76 hm²
Gross floor area:	93,506 m²
F.A.R.:	0.84

珠海香洲淇澳游艇会
Qi'ao Yacht Club, Xiangzhou, Zhuhai

地点：中国，珠海
业主：珠海金山软件股份有限公司
设计范围：方案设计
设计时间：2013

Location: Zhuhai, China
Client: Zhuhai Kingsoft Co. Ltd.
Design scope: Proposal Design
Design time: 2013

本项目位于珠海市淇澳岛南芒湾北侧，是一所度假主题俱乐部，包括陆地会所和海上游艇停泊区域。

会所的设计吸纳了航海元素，以展现"自由、自然、挑战"的精神。内部空间的设计以舒适体验为核心，功能区的布局在充分考虑海面景观最大化的前提下进行。朝海延展面的最大化保证了公共空间，例如，办公室、餐厅、水疗室、泳池的景观优势。同时，所有的贵宾房均可欣赏到海景。

The project is located in the northern part of Nanmang Bay, Qi'ao Island, Zhuhai. The project provides both a club on the ground and a yacht docking area on the sea to create a holiday-themed yacht club.

The club design incorporates nautical elements to promote the "free, nature and challenge" spirits. The inner space design mainly focuses on the comfort experience. The priority of the layout of functional areas is to maximize the magnificent ocean view. The curved facade towards the sea is able to accommodate the public space such as offices, restaurants, a spa and a swimming pool, at the same time providing ocean views to every VIP room.

总用地面积：	1.48 hm²
总建筑面积：	14 820 m²
容积率：	1.0

Site area:	1.48 hm²
Gross floor area:	14,820 m²
F.A.R.:	1.0

珠海横琴天沐河游艇会
Tianmu River Yacht Club, Hengqin, Zhuhai

地点：中国，珠海
业主：东西汇（横琴）发展有限公司
设计范围：方案设计
设计时间：2020

Location: Zhuhai, China
Client: WE Park (Hengqin) Development Co., Ltd.
Design scope: Proposal Design
Design time: 2020

本项目北邻天沐河，东邻东西汇综合体，是天沐河边的一道靓丽风景线。

设计采用了三条海豚跃出水面的意向。游艇会采用钢结构顶棚，强调线条的流畅性，并与现代建筑材料有机结合，配以银色和白色，形成一种精致的空间及立面效果。

游艇会主要设置有游艇停泊、餐饮娱乐、红酒和雪茄吧及水疗等功能，为游艇爱好者提供了一种时尚个性的休闲生活方式。

This project faces Tianmu River on the north and the Complex of WE Park on the east, which is a beautiful scenic by the Tianmu River.

The design adopts the conceptual image of three dolphins jumping out of the water. The yacht club is built with steel structure roofs, emphasizes the fluency of lines, and is organically combined with modern building materials, with silver and white colours to form a delicate space and facade effect.

The yacht club is mainly equipped with yacht mooring, dining and entertainment, wine and cigar bar, spa and other functions, providing a stylish and individual leisure lifestyle for the yacht lovers.

总用地面积：	1.01 hm^2
总建筑面积：	6 000 m^2
容积率：	0.6

Site area:	1.01 hm^2
Gross floor area:	6,000 m^2
F.A.R.:	0.6

珠海横琴东西汇大剧院
WE Park Grand Theatre, Hengqin, Zhuhai

地点：中国，珠海
业主：东西汇（横琴）发展有限公司
设计范围：方案设计、扩初设计、室内设计、剧目设计
设计时间：2019

Location: Zhuhai, China
Client: WE Park (Hengqin) Development Co., Ltd.
Design scope: Proposal Design, Design Development, Interior Design, Musical Design
Design time: 2019

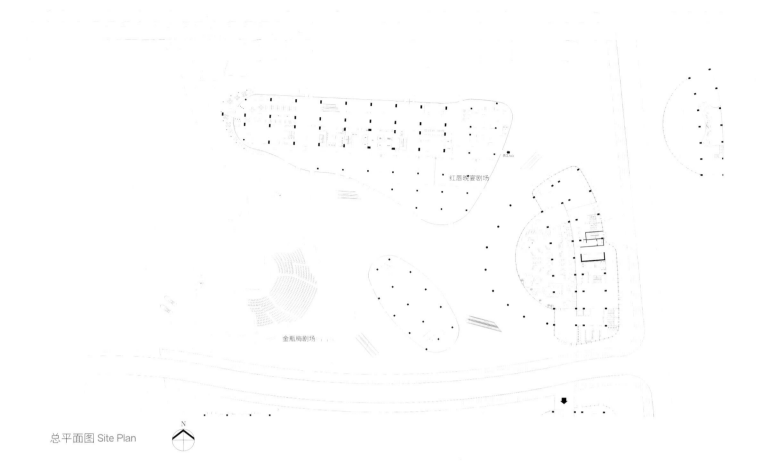

总平面图 Site Plan

东西汇大剧院主要由沉浸式剧场、音乐餐厅、600座晚宴剧场和实景剧场四部分组成。

沉浸式剧场是由能够自由升降的水舞台，以及20世纪30年代风格的澳门实景建筑、贵宾室和观众席组成。澳门实景建筑可同时作为表演场景及酒店客房。沉浸式剧场在未来可以改造成1 500座标准剧场。

客人在音乐餐厅中可以欣赏到邓丽君的经典老歌，仿佛回到了20世纪80年代。

晚宴剧场将引入纽约百老汇等多种表演形式，为客人提供精彩的表演。

实景剧场把客人拉回到宋代，通过瓷器、书画等展览和商业业态，为之提供传统戏曲的享受。

沉浸式剧场面积：	10 000 m²
沉浸式剧场容量：	1 500 人
音乐餐厅面积：	730 m²
音乐餐厅容量：	300 人
晚宴剧场面积：	5 000 m²
晚宴剧场容量：	600 人
实景剧场面积：	1 000 m²
实景剧场容量：	100 人

WE Park Grand Theatre is mainly composed of four sections: an immersive theatre, a music restaurant, a 600-seat dinner theatre and a live theatre.

The immersive theatre is composed of a water stage that can be raised and lowered freely, as well as the live scene of Macao in the 1930s, VIP rooms, and the auditorium. The live scene of Macao can be a performing area as well as guest rooms. The immersive theatre can be renovated into a standard theatre with 1,500 seats in the future.

The music restaurant is where the guests could admire Teresa Teng's classic songs, which bring people back to the 1980s.

The dinner theatre will introduce a variety of performance forms such as the Broadway show of New York, to provide guests with passionate performances.

The live theatre will bring guests back to the Song Dynasty. It provides the enjoyment of traditional operas through exhibitions of porcelain, calligraphy and painting as well as other commercial formats.

Area of the immersive theatre:	10,000 m²
Capacity of the immersive theatre:	1,500
Area of the music restaurant:	730 m²
Capacity of the music restaurant:	300
Area of the dinner theatre:	5,000 m²
Capacity of the dinner theatre:	600
Area of the live theatre:	1,000 m²
Capacity of the live theatre:	100

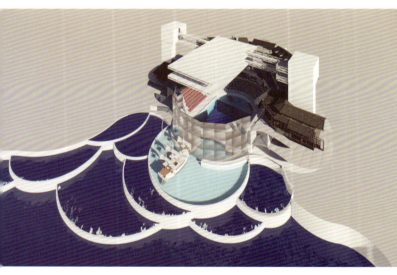

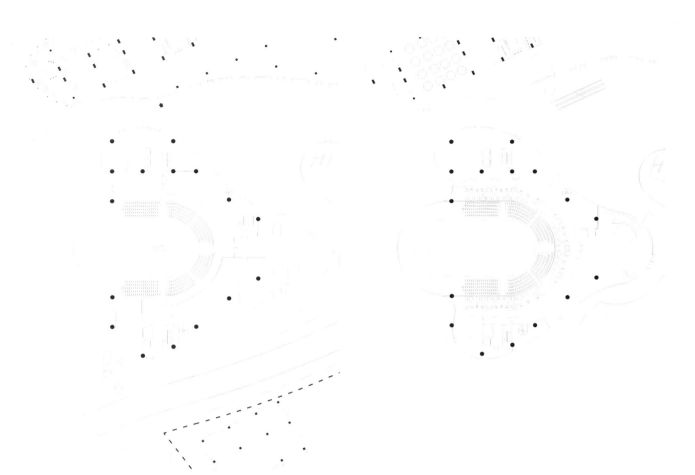

沉浸式剧场一层平面图　1st Floor Plan of the Immersive Theatre　　沉浸式剧场二层平面图　2nd Floor Plan of the Immersive Theatre

沉浸式剧场三层平面图　3rd Floor Plan of the Immersive Theatre　　沉浸式剧场四层平面图 4th Floor Plan of the Immersive Theatre

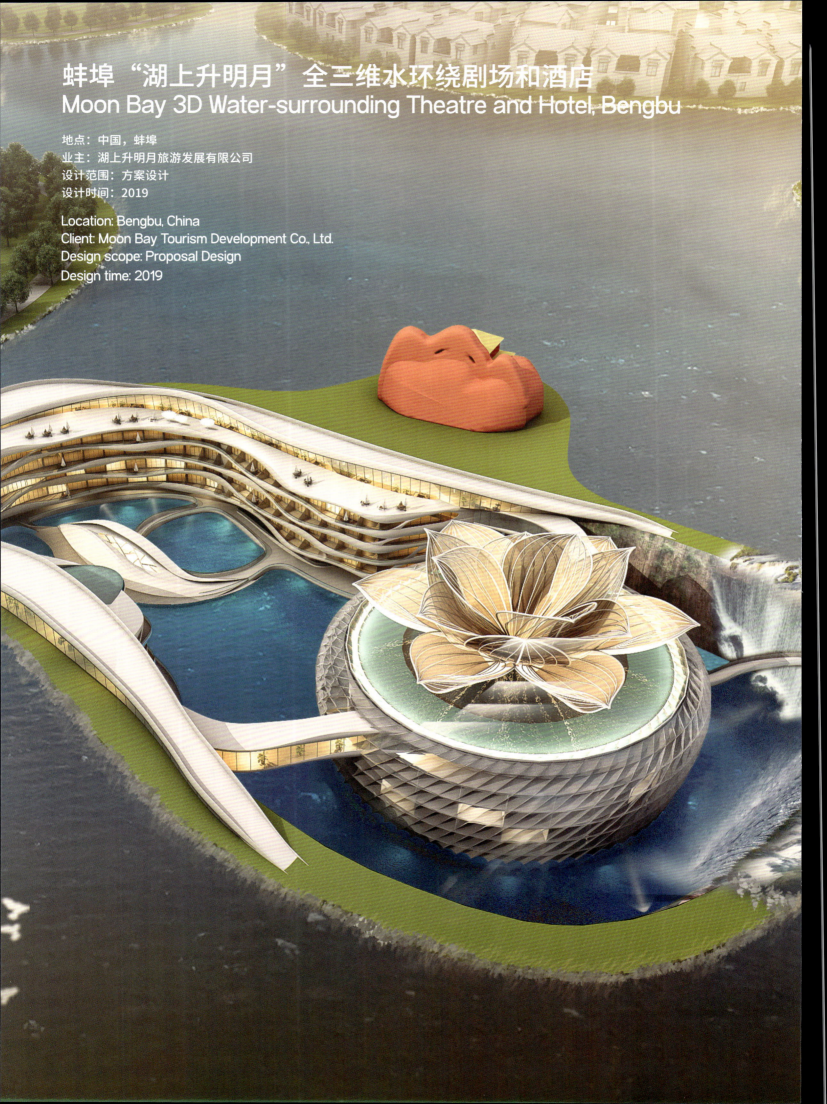

蚌埠"湖上升明月"全三维水环绕剧场和酒店
Moon Bay 3D Water-surrounding Theatre and Hotel, Bengbu

地点：中国，蚌埠
业主：湖上升明月旅游发展有限公司
设计范围：方案设计
设计时间：2019

Location: Bengbu, China
Client: Moon Bay Tourism Development Co., Ltd.
Design scope: Proposal Design
Design time: 2019

总平面图　Site Plan

全三维水环绕剧场位于湖上升明月景区内部。项目基地是一个南北长140 m、东西长70 m、深20 m的不规则深坑。设计方案创造性地设计了全三维水环绕剧场，以莲花为主要形态概念。

项目主体是拥有360°水中舞台的水环绕剧场，并全方位环绕观众厅布置，让全体观众都可以欣赏到水下表演。演员通过莲花下的屋顶进入水下舞台。项目还拥有约88个水下房间和一间水下餐厅。深坑一侧的崖壁上还设置了一整面LED显示屏。深坑另一侧是崖壁酒店，共100间客房，所有的客房都面向中心的水池，拥有良好的景观。

项目在设计之初就将水下表演纳入设计范围。在设计阶段，设计将建筑和艺术表演充分结合，优化演出流程和剧场管理。剧情也充分结合建筑形式，最终为观众呈现出精彩的演出内容。

The 3D water-surrounding theatre is located inside the Moon Bay scenic area. The project site is an irregular deep pit with a length of 140 m from north to south, 70 m from east to west and a depth of 20 m. The design scheme creatively designs the 3D water-surrounding theatre that generated from the image of a lotus.

The main part of the project is a water-surrounding theatre, with a 360° immersive underwater stage and is arranged around the audience hall, so that the audience can enjoy the underwater performances. The actors can enter the underwater stage through the roof under the lotus. The project also has about 88 rooms surrounded by water and an underwater restaurant. There is also a full LED display on the cliff. On the other side of the deep pit is the cliff hotel with 100 rooms in total. All rooms have an excellent view, facing the central pool.

The project incorporated the underwater performances into the design scope at the beginning stage, and integrated architecture and artistic performances during the design process. The design optimized the performing process and theatre management, and fully integrated the plots with the architectural form to present wonderful performances for the audience.

总用地面积：	1.63 hm²
总建筑面积：	18 500 m²
水下剧场面积：	2 000 m²（300座）
崖壁酒店面积：	6 705.04 m²（100间客房）
水下酒店面积：	6 332.56 m²（88间客房）
水下餐厅面积：	801.06 m²

Site area:	1.63 hm²
Gross floor area:	18,500 m²
Underwater theatre area:	2,000 m² (300 seats)
Cliff hotel area:	6,705.04 m² (100 rooms)
Underwater hotel area:	6,332.56 m² (88 rooms)
Underwater restaurant area:	801.06 m²

上海黄浦上海文化广场KFS生态箱剧场
The Music Box by KFS at Shanghai Cultural Square, Huangpu, Shanghai

地点：中国，上海
业主：上海文化广场剧院管理有限公司
设计范围：方案设计、施工图设计
设计时间：2014
建成时间：2014

Location: Shanghai, China
Client: Shanghai Culture Square Theatre Management Co., Ltd.
Design scope: Proposal Design, Construction Documentation
Design time: 2014
Completion time: 2014

KFS生态箱剧场由集装箱改建，有别于传统的镜框式舞台，剧场舞台环绕座位设置，设观众席300个，全方位展现小剧场戏剧作品的先锋艺术特性。观众可近距离接触与自然环境相融合的舞台。

在观众席方面，360°可旋转座椅可欣赏到四面舞台的表演效果，贯彻生态环保的创意理念。观众置身其中，以一种创新模式解构戏剧，拉近普通观众与舞台演员的距离，利用半封闭的舞台形式，将舞台充分融入自然环境，让艺术真正成为一种生活常态。

对于户外舞台的开发，上海文化广场旨在通过每年100场先锋的、试验性的文化表演、艺术活动等项目运作，吸引18至40岁的观众群体，提供有特色的观演体验及互动活动，演艺内容力求时尚、现代、多元。

The Music Box by KFS is rebuilt with containers, which is different from the traditional frame stage. The theatre has an auditorium of 300 seats, surrounded by stages on four sides. It shows fully the pioneering artistic characteristics of drama works in small theatres. The audiences are close to the stages which are integrated with the natural environment.

In terms of the auditorium, 360° rotatable seats provide the view for performance on all four stages, and implement the creative concept of ecological and environmental protection. The theatre deconstructs the drama innovatively, narrowing the distance between the ordinary audiences and the actors by using the semi-enclosed stage form to fully integrate the stage into the natural environment and making art a norm of life.

For the development of outdoor stages, Culture Square aims to attract audiences aged 18 to 40 through the operation of 100 avant-garde, experimental cultural performances, artistic activities and other projects every year and provides distinctive experience and interactive activities whose contents strive for fashion, modernity and diversity.

| 总用地面积： | 0.0324 hm² |
| 观众容量： | 300人 |

| Site area: | 0.0324 hm² |
| Audience capacity: | 300 |

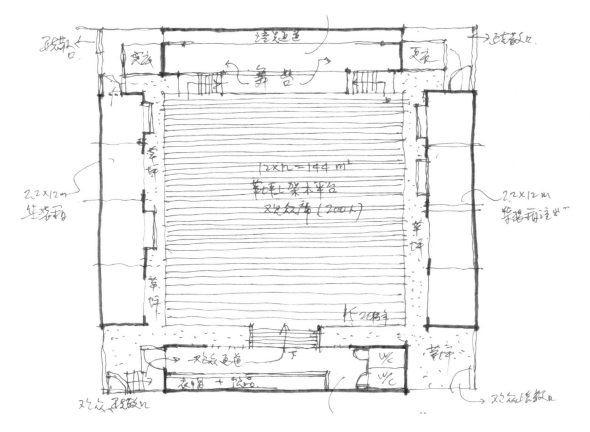

设计草图1 Sketch 1

设计草图2 Sketch 2

中华人民共和国驻外某总领事馆
A Consulate General of the People's Republic of China

地点：加拿大
业主：中华人民共和国外交部
设计范围：方案设计
设计时间：2019

Location: Canada
Client: Ministry of Foreign Affairs of the People's Republic of China
Design scope: Proposal Design
Design time: 2019

总领事馆坐落在山体和自然区一块面积为7307.38 m²的土地上，可欣赏到市区、河流的全景。

建筑的意向结合了周围的山地，呼应了周边环境。同时，这种抽象的屋顶意向也体现了东方元素和当地元素的融合。

总建筑面积：　　9 016 m²

The Consulate General is located on the 7,307.38 m² land in the historical and natural area of Mountain. It can enjoy the panoramic view of downtown and river .

The intention of the building combines with the surrounding mountains echoes the surrounding environment. Meanwhile, the abstract roof also reflects the fusion of oriental elements and local elements.

Gross floor area:　　　9,016 m²

哈尔滨道里爱建新城交银大厦
BCM Tower in Aijian New City, Daoli, Harbin

地点：中国，哈尔滨
业主：上海爱建股份有限公司 / 哈尔滨爱达投资置业有限公司
设计范围：方案设计、扩初设计
设计时间：2003
建成时间：2007

Location: Harbin, China
Client: Shanghai Aijian Co., Ltd. / Harbin Aida Investment Real Estate Co., Ltd.
Design scope: Proposal Design, Design Development
Design time: 2003
Completion time: 2007

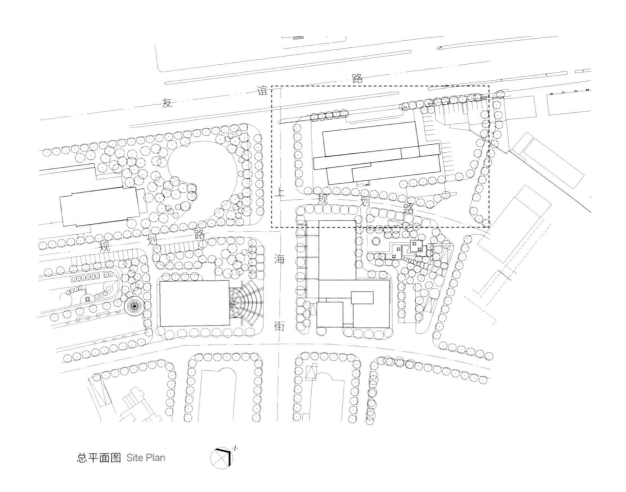

总平面图 Site Plan

本项目位于友谊路和上海街的交叉路口，作为KFS设计的哈尔滨爱建新城的一部分，它是交通银行以办公为主的综合性大厦，可以远眺松花江。

设计的目标是在满足功能需求的同时，营造一栋既与冰城文脉相协调，又不失信息时代独特品格的建筑。建筑西侧宛如松花江畔一挂瀑布般的玻璃跌落体为设计的重点，亦为信息时代品格的象征，构成符合业主精神特质的设计主题。树木、植被和瀑布等自然要素通过架空、渗透等手段，使环境延伸入建筑，又使建筑融入城市环境，一起共生共长。设计用现代的材料和独特的手段演绎建筑的宁静、优雅，使之成为哈尔滨最独特的建筑之一。

The project is located at the intersection of Youyi Road and Shanghai Street. As part of the Harbin Aijian New City designed by KFS, it is a comprehensive office building of the Bank of Communications, overlooking the Songhua River.

The goal of the design is to meet the functional requirements when creating a building that is in harmony with the ice-city culture and to possess unique character of the information age. The glass stack on the west side of the building resembles a hanging waterfall beside the Songhua River, and it is also a symbol of the character of the information age, constituting a design theme that conforms to the spiritual characteristics of the owner. Natural elements such as trees, vegetation, and waterfalls extend the environment into the building through means such as lifting and infiltration, and integrate the building into the urban environment to co-exist and grow together. The design uses modern materials and unique means to interpret the tranquility and elegance of the building, making it one of the most unique buildings in Harbin.

总用地面积:	0.55 hm²
总建筑面积:	25 000 m²
容积率:	4.6

Site area:	0.55 hm²
Gross floor area:	25,000 m²
F.A.R.:	4.6

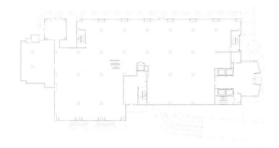

一层平面图　1st Floor Plan

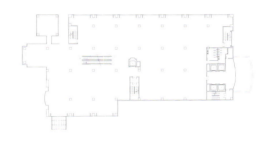

二层平面图　2nd Floor Plan

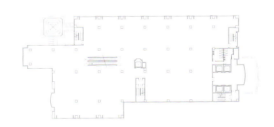

三层平面图　3rd Floor Plan

标准层平面图　Typical Floor Plan

上海长宁舜元大厦
Shunyuan Office Tower, Changning, Shanghai

地点：中国，上海
业主：上海北大青鸟企业发展有限公司
设计范围：方案设计、扩初设计
设计时间：2002
建成时间：2008

Location: Shanghai, China
Client: APTECH Enterprise Development Inc.
Design scope: Proposal Design, Design Development
Design time: 2002
Completion time: 2008

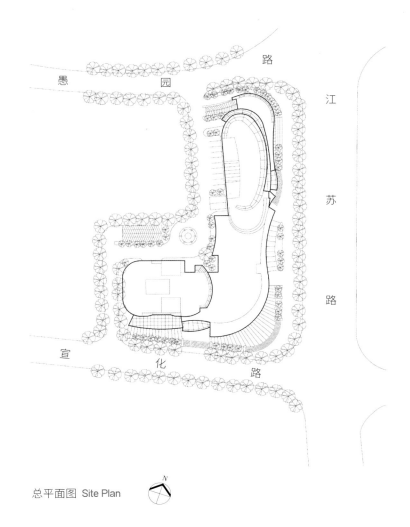

总平面图 Site Plan

本项目位于上海市长宁区江苏路、愚园路与宣化路交会处，是一栋混合功能的豪华大楼，包括豪华酒店、服务公寓、专业办公空间和商业设施。

时尚与简洁是这一地区重要地标的主要设计标准。该标准通过细节的对比和呼应体现在立面的表达中。墨守成规、单调乏味的连续立面被有意识地舍去。该设计诠释了色彩的使用及细节的智慧。

办公楼结合地形采用了偏心筒布置，这样不但使办公面积经济有效，而且又不失外形的活泼流畅。公寓楼采用单元平面拼接式，优美的轮廓与办公楼相得益彰。三层的裙楼以商业为主，建筑形象呈现出弯曲的流线型，在周围的环境中别具一格。通透的玻璃、光滑的铝板、细致的不锈钢装饰使建筑整体流畅、优美。

This project is located at the intersection of Jiangsu Road, Yuyuan Road and Xuanhua Road in Changning District, Shanghai. It is a mixed-function luxury building, including a luxury hotel, service apartments, a professional office space and commercial facilities.

Style and simplicity are the main design criteria for the high profile landmark in the community. Both criteria find their expressions in the facade through the contrast and consistency of detailing. The stereotypical, monotonous and continuous facade has been consciously avoided. The design underlines the use of and intelligent of detailing.

The office building adopts an eccentric tube layout in combination with the terrain, which not only makes the office area economical and effective, but also has a lively and smooth appearance. The apartment building adopts plane splicing style, and the beautiful outline complements the office building. The three-storey podium is mainly commercial, and the architectural image presents a curved streamline, which is unique in the surrounding environment. Transparent glass, smooth aluminum plates and delicate stainless steel decorations make the building complex smooth and beautiful.

总用地面积:	0.7 hm²
总建筑面积:	45 000 m²
容积率:	6.3

Site area:	0.7 hm²
Gross floor area:	45,000 m²
F.A.R.:	6.3

上海浦东世纪大道长泰国际金融大厦
Changtai Office Tower on Century Boulevard, Pudong, Shanghai

地点：中国，上海
业主：长甲集团国际控股有限公司
设计范围：方案设计、扩初设计
设计时间：2002
建成时间：2008

Location: Shanghai, China
Client: Changjia Group Intl. Holding Ltd.
Design scope: Proposal Design, Design Development
Design time: 2002
Completion time: 2008

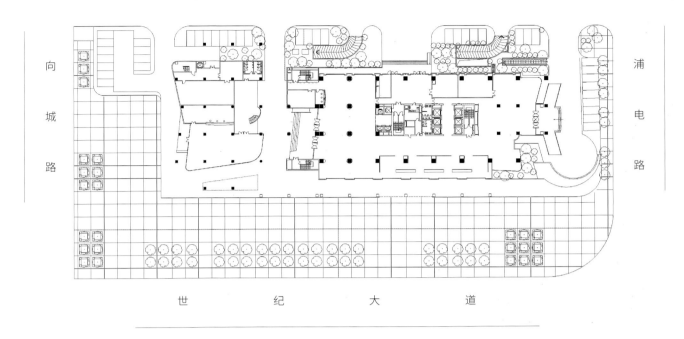

向城路

浦电路

世 纪 大 道

总平面图 Site Plan

本项目位于上海浦东世纪大道，北面为期货交易大厦，南面为世纪大道，是一座以办公为主体，兼具会议、娱乐、银行、保险等功能的综合性大厦。100 m的限高造就了新古典主义设计。立面细节采用石头与玻璃，创造出"独特的建筑宣言"。建筑外立面呈现出坚实、稳固、豪华的外形。

在主体办公楼内部，每四层均设有一个通高的空中花园，人们透过它可欣赏到世纪大道壮丽的景色。空中花园加强了不同楼层的联系，为员工休息提供了宁静、优雅的场所。为了在狭窄的基地上创造一种优雅舒适的外部环境，建筑底部被设计成开放式的室外广场，面向与之相邻的公共绿地，并向世纪大道敞开。同时，设计结合世纪大道已有的景观绿化，设置了室外广场及绿地、树木、雕塑等小品，使整个基地与世纪大道的景观相融合。

The project is located on Century Avenue in Pudong, Shanghai, with the Futures Trading Building to the north and Century Avenue to the south. It is a comprehensive building with offices as the main body and also for conferences, entertainment, banking, insurance and other purposes. The height limit of 100 m creates a neoclassical design. The facade details use stone and glass to create a "unique architectural manifesto". The facade of the building presents a solid, stable and luxurious form.

Inside the main office building, every fourth floor is equipped with a full-height sky garden, through which people can enjoy the magnificent view of Century Avenue. The sky gardens strengthen the connection between different floors, providing quiet and elegant places for employees to relax. In order to create an elegant and comfortable external environment on the narrow base, the bottom of the building is designed as an open outdoor plaza, facing the adjacent public green space and opening to Century Avenue. Meanwhile, combined with the existing landscape of Century Avenue, outdoor square and green space, trees, sculptures and other sketches are set, so that the entire site is integrated with the Century Avenue landscape.

总用地面积:	0.9 hm²
总建筑面积:	57 000 m²
容积率:	5.0

Site area:	0.9 hm²
Gross floor area:	57,000 m²
F.A.R.:	5.0

上海浦东东晶国际办公大厦
Dongjing International Office Complex, Pudong, Shanghai

地点：中国，上海
业主：上海东道置业有限公司
设计范围：方案设计、扩初设计
设计时间：2002
建成时间：2005

Location: Shanghai, China
Client: Shanghai Dongdao Real Estate Inc.
Design scope: Proposal Design, Design Development
Design time: 2002
Completion time: 2005

总平面图 Site Plan

本项目位于源深路以东，浦东大道以南。地块为东西走向的狭长地形，办公楼总高为24层（不含屋顶设备层）。

这座100 m高的地标建筑以其独特、优雅的形象成为建筑的表达。夜晚璀璨的灯光与建筑为城市增光添彩，犹如披着华丽的外衣。办公楼平面为方形，标准层建筑面积约为1 000 m²，适应自由灵活的分隔。每层分别设置空调机组，满足小型办公单位的需要。

总用地面积：	1.6 hm²
总建筑面积：	66 000 m²
容积率：	4.0

The project is situated in the east of Yuanshen Road, and the south of Pudong Avenue. The site is a narrow rectangle from east to west. The office building is 24 floors (excluding the roof equipment floor).

The 100 m high landmark is an architectural statement with its unique and elegant image. Ornate neon lights at night dazzle the city with the building, seemingly draped in a sparkling trendy outfit. The office building is a square shape in plan, and the standard floor area is about 1,000 m². It is suitable for free and flexible separation. Each floor is equipped with air conditioning units to meet the needs of small office units.

Site area:	1.6 hm²
Gross floor area:	66,000 m²
F.A.R.:	4.0

哈尔滨道里爱建新城SOHO
Aijian New City SOHO, Daoli, Harbin

地点：中国，哈尔滨
业主：上海爱建股份有限公司 / 哈尔滨爱达投资置业有限公司
设计范围：方案设计、扩初设计
设计时间：2002
建成时间：2008

Location: Harbin, China
Client: Shanghai Aijian Co., Ltd. / Harbin Aida Investment Real Estate Co., Ltd.
Design scope: Proposal Design, Design Development
Design time: 2002
Completion time: 2008

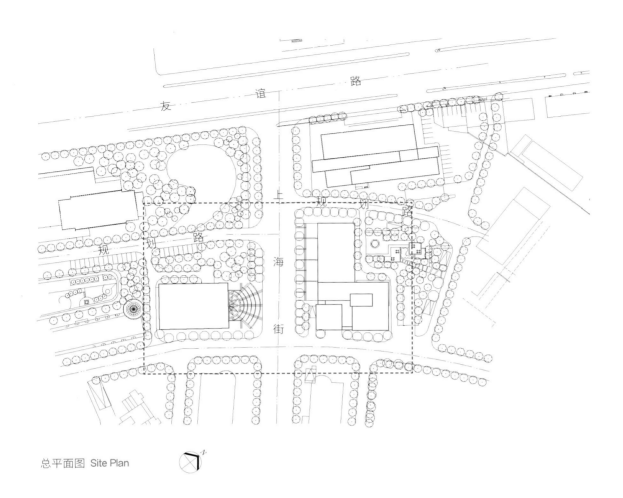

总平面图 Site Plan

整个爱建新城以中心广场为核心，呈环状布置。本项目属于其西北片区。两栋SOHO建筑隔着上海街相望，形成爱建新城的门户建筑和标志性形象。

两栋SOHO建筑主要采用玻璃及铝板材质，与沿街已建及保留建筑相呼应，挺拔、简练的体型塑造手法使其成为本街区的新地标。考虑当地特点和功能性质，色彩上更多运用中等明度及中性色调，使整体形象既统一又俏丽，让人们四季生活在生动活泼的氛围中。

The entire Aijian New City has a central square as its core and is arranged in a ring shape. This project belongs to its northwest area. The two SOHO buildings face each other across Shanghai Street, forming the gateway building and iconic image of Aijian New City.

The two SOHO buildings are mainly made of glass and aluminum panels, which echo the existing and preserved buildings along the street. The upright and concise body shaping techniques make them the new landmarks of the district. Taking into account the local characteristics and functions, more use of medium lightness and neutral tones in the colour makes the whole image uniform and beautiful, and allows people to live in a lively atmosphere in all seasons.

总用地面积：	0.9 hm²
总建筑面积：	50 000 m²
容积率：	5.8

Site area:	0.9 hm²
Gross floor area:	50,000 m²
F.A.R.:	5.8

珠海横琴口岸综合服务中心
Comprehensive Service Centre of Hengqin Port, Hengqin, Zhuhai

地点：中国，珠海
业主：珠海大横琴置业有限公司
设计范围：方案设计
设计时间：2019

Location: Zhuhai, China
Client: Zhuhai Dahengqin Real Estate Co., Ltd.
Design scope: Proposal Design
Design time: 2019

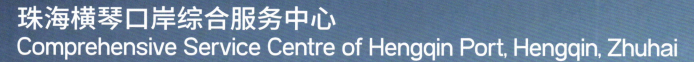

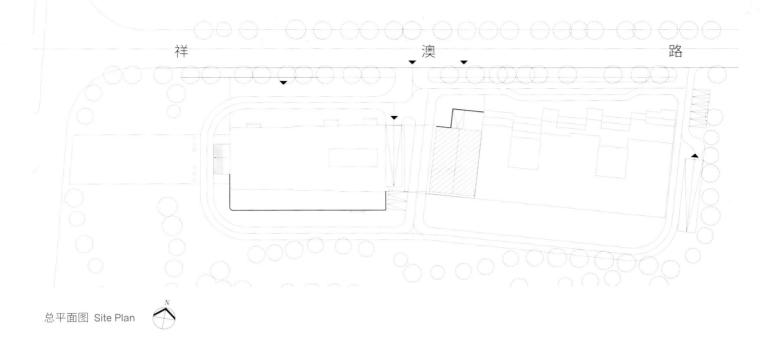

祥　　　　　澳　　　　　路

总平面图 Site Plan

本项目意在打造横琴口岸的综合服务中心，同时也将成为该地区的地标建筑。

建筑塔楼以及裙楼通过跌落的形态，营造出层层叠叠的屋顶花园空间，增加了办公人员以及住户的户外活动空间。裙楼顶部设置有空中水公园以及屋顶餐厅，为办公人员以及住户提供休闲和娱乐的绿色空间。

本项目的核心是一栋90 m高的政务中心塔楼，以及一栋150 m高的公寓塔楼。公寓塔楼以小户型为主，易于居住，并且在1至8层设置了游泳池、健身房、咖啡馆等大量配套设施。这些空间作为居住功能的延伸满足了住户的全面需求。

横琴口岸综合服务中心将充分发挥横琴的区位、环境和政策优势，承接口岸服务配套功能，为横琴口岸提供"一站式"综合性政务解决方案。横琴口岸综合服务中心将成为横琴重要的展示形象。

The project is aiming to build a Comprehensive Service Centre of Hengqin Port, and will become a landmark in the region.

The towers and podium of the building create a layered roof garden space through setting back level by level, adding the outdoor activity space for the office workers and the residents. At the top of the podium, there is a sky water park and a rooftop restaurant, which provide a green space for the office workers and the residents to relax and entertain.

The core of the project is a 90 m high government affairs centre tower and a 150 m high apartment tower. The apartment tower is mainly composed of easy-living small-sized condos and a large number of supporting facilities such as the swimming pool, the gym, the cafe and etc., which are located on the first to the eighth floor. These spaces serve as an extension of living functions, ensuring the general needs of the residents are met.

The Comprehensive Service Centre of Hengqin Port will take full advantage of location, environment and policy of Hengqin. It will undertake the supporting functions of port services, and provide a "one-stop" comprehensive government affairs solution for Hengqin Port. The Comprehensive Service Centre of Hengqin Port will become an important image of Hengqin.

总用地面积：	0.84 hm²
总建筑面积：	76 560.16 m²
容积率：	6.79
建筑密度：	39.76%

Site area:	0.84 hm²
Gross floor area:	76,560.16 m²
F.A.R.:	6.79
Building density:	39.76%

珠海横琴码头综合交通开发
Hengqin Terminal Complex, Hengqin, Zhuhai

地点：中国，珠海
业主：珠海大横琴泛旅游发展有限公司
设计范围：方案设计
设计时间：2019

Location: Zhuhai, China
Client: Zhuhai Dahengqin Pan Tourism Development Co., Ltd.
Design scope: Proposal Design
Design time: 2019

总平面图 Site Plan

环
横琴码头
宝
岛
东
兴
路
路

N

本项目位于横琴东南角，与澳门氹仔岛隔海相望。基地东侧是现状码头，为西班牙风格；西侧是横琴镇商业筑建，为南洋风格；北侧是公园及立体停车楼，为现代风格；南侧是简易棚状海鲜店：周边建筑风格多样化。基地位于横琴本岛东侧休闲长廊带，是本岛和南岛之间的重要功能联结点。

方案构思来源于漂浮于水面之上的落叶，设计概念是"一叶扁舟，带我去诗和远方；一叶扁舟，载我去想去的地方"。

参照周边的各式建筑形式：西班牙风格、南洋风格、现代风格等，设计的策略为师法自然，和谐包容。设计使建筑与自然和谐统一，用绿叶象征建筑形式。

The project is located in the southeast corner of Hengqin, across the sea from Taipa Island, Macao. The east side of the site is the current pier in Spanish style; the west side of the site is the commercial building in Nanyang style; the north side of the site is the modern style park and car parking building; the south side of the site is street hawkers. The surrounding architectural styles are diverse. The site is located in the leisure promenade belt on the east side of Hengqin main island and is an important connection point between the main island and the south island.

The design idea comes from the fallen leaves floating on the water. The design concept is: "One leaf flat boat, brings me to the poetry and the distance. One leaf flat boat, takes me to the place I want to go".

When referring to the surrounding architectural forms of Spanish style, Nanyang style, and modern style, etc., the design strategy is to imitate nature with harmony and tolerance. The design harmonizes and unifies the building with nature and uses green leaves to symbolize the architectural form.

| 总用地面积： | 1.1 hm² |
| 总建筑面积： | 14 084 m² |

| Site area: | 1.1 hm² |
| Gross floor area: | 14,084m² |

珠海横琴50#地块华发综合体
Lot 50# Huafa Complex, Hengqin, Zhuhai

地点：中国，珠海
业主：珠海华发房地产开发有限公司
设计范围：方案设计、扩初设计
设计时间：2018

Location: Zhuhai, China
Client: Zhuhai Huafa Real Estate Development Co., Ltd.
Design scope: Proposal Design, Design Development
Design time: 2018

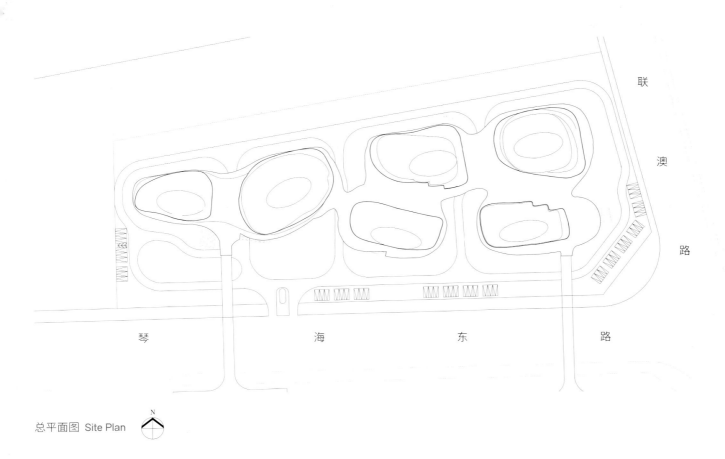

总平面图 Site Plan

联
澳
路

琴　海　东　路

本项目位于珠海横琴的十字门中央商务区，与澳门隔海相望。本地块与周边的其他地块一同营造出充满活力的办公、商业居住综合体。

基地紧邻滨水空间，建筑方案犹如粤港澳大湾区的一朵精美浪花，充满动感与活力。设计受到珠海滨水文化的启发，裙房的造型令人联想起流水的形态，带来漂浮的感觉，进而营造出各具特点的商业氛围与社区空间。此外，裙房房顶的户外空间与澳门塔形成对景。

The project is located in Shizimen CBD of Hengqin, Zhuhai. There are beautiful views across the sea, connecting with Macao. Together with the surrounding sites, this site contributes to a vital complex of office, commercial and residential spaces.

This site is adjacent to the waterfront, and the building plan is like a delicate splash in the Guangdong-Hong Kong-Macao Greater Bay Area; an area full of movement and vitality. The design is inspired by the Zhuhai coastal culture. The shape of the podium is reminiscent of flowing water and contributes to the feeling of floating, thus creates characteristics which stimulate the commercial atmosphere and community space. In addition, the podium rooftop uses its outdoor space as the opposing view of the Macao Tower.

总用地面积：	2.5 hm²
总建筑面积：	51 800 m²
容积率：	2.0

Site area:	2.5 hm²
Gross floor area:	51,800 m²
F.A.R.:	2.0

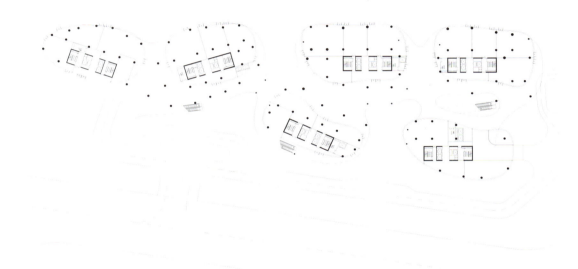

一层平面图 1st Floor Plan

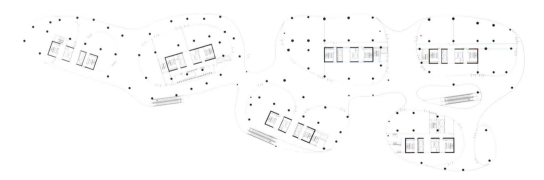

三层平面图 3rd Floor Plan

珠海横琴51#、52#地块办公综合体
Lot 51#, 52# Office Complex, Hengqin, Zhuhai

地点：中国，珠海
业主：圣马丁国际控股有限公司
设计范围：方案设计
设计时间：2018

Location: Zhuhai, China
Client: Sandmartin International Holdings Limited
Design scope: Proposal Design
Design time: 2018

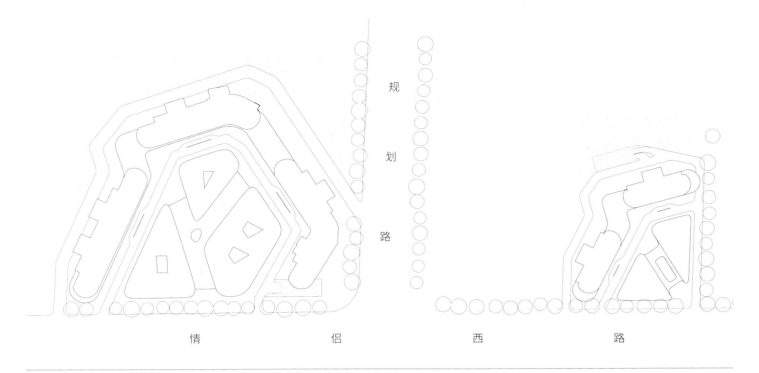

总平面图 Site Plan

规

划

路

情　侣　西　路

　　项目基地位于珠海横琴，与小横琴岛隔江相望。基地现状平整，基础设施完善。

　　基地景观资源优越，坐拥一线江景，南望大小横琴山，北望竹仙公园，东望横琴大桥和澳门。基地南濒马骝洲水道，距国际航线大西水道4 n mile；西接磨刀门水道，与珠海西区一衣带水；东与澳门一桥相通，距香港41 n mile。周边有多个机场，地理位置极为优越。

　　方案通过优美的曲线呼应江景，以柔和的体量来应对周围优越的景观环境。地上1至3层为商业空间，建筑体态结合地形，静谧自然。地上4至49层为SOHO办公空间，景观开阔。建筑布局使得每个房间都有优美的江景，大部分单元可以把澳门景色纳入视野。

The site is located in Hengqin, Zhuhai, across the river from Xiaohengqin Island. The current status of the site is flat and the infrastructure is fully complete.

The site has abundant landscape resources and fantastic river views, facing south to Dahengqin Mountain and Xiaohengqin Mountain, facing north to Zhuxian Park, facing east to Hengqin Bridge and Macao. It borders Maliuzhou Waterway in the south, and is 4 n mile away from the international route—Great West Waterway. It is connected to Modaomen Waterway in the west and connects with Zhuhai West District. The site is connected with Macao by a bridge in the east and is 41 n mile away from Hong Kong. It is also surrounded by multiple airports in the vicinity. Thus, the site has a prime geographical location.

The scheme echoes the river views through beautiful curves, and responds to the superior surrounding landscape environment with a soft volume. The 1st to the floors above ground are commercial space. The form is combined with the terrain and is quiet and natural. The 4th to the 49th floors are SOHO office space that have open landscape views. The layout of the building allows each room to have a beautiful river view, and many of the units have pleasant views of Macao.

51#地块

总用地面积：	1.99 hm²
总建筑面积：	153 752 m²
容积率：	6.0

52#地块

总用地面积：	0.76 hm²
总建筑面积：	49 846 m²
容积率：	5.0

Lot 51#

Site area:	1.99 hm²
Gross floor area:	153,752 m²
F.A.R.:	6.0

Lot 52#

Site area:	0.76 hm²
Gross floor area:	49,846 m²
F.A.R.:	5.0

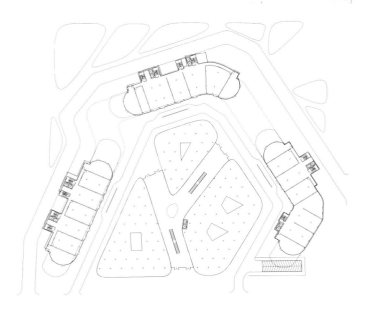
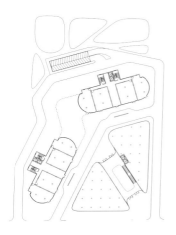

一层平面图 1st Floor Plan

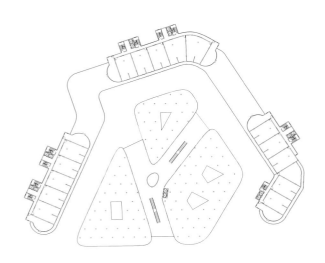
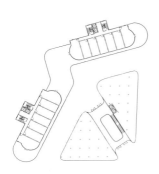

二层平面图 2nd Floor Plan

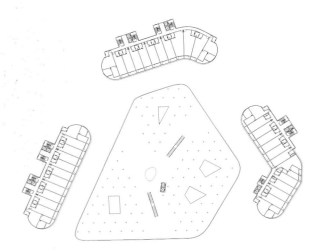
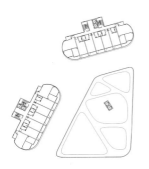

标准层平面图 Typical Floor Plan

珠海横琴华发世界汇38#地块
Lot 38# of Huafa World Centre, Hengqin, Zhuhai

地点：中国，珠海
业主：珠海华发房地产开发有限公司
设计范围：方案设计
设计时间：2018

Location: Zhuhai, China
Client: Zhuhai Huafa Real Estate Development Co., Ltd.
Design scope: Proposal Design
Design time: 2018

总平面图 Site Plan

38#地块主要用于LOFT类公寓、酒店和商业空间等。

整体概念来源于中国古典窗棂以及立体城市，结合周边建筑及景观资源情况合理布局，充分利用、挖掘、优化景观资源；利用及拓展现有交通资源及体系，合理组织不同业态的交通流线，构建项目高效便捷的交通系统。设计充分考虑标准化设计，采用节能环保技术，保证材料的长期运营，并在其服务寿命内取得最大的经济效益，以最少的能耗和运行费用提供舒适的内部环境。

Lot 38# are mainly used as loft apartments, hotel and commercial space.

The overall concept is derived from Chinese classical window lattices and three-dimensional cities, rationally arranged with the surrounding buildings and the landscape resources. It fully utilized, excavated and optimized the landscape resources. It used and expanded the existing transportation resources and systems to rationally organize the traffic flow of different business formats, constructing an efficient and convenient transportation system. The design fully considers the standardized design and adopts energy-saving and environmental protection technology to ensure the long-term operation of the materials, and to obtain the maximum economic benefits within the service life. It is also the goal to provide a comfortable internal environment with the least energy consumption and operating costs.

总用地面积：	3.2 hm²
总建筑面积：	196 000 m²
容积率：	6.12

Site area:	3.2 hm²
Gross floor area:	196,000 m²
F.A.R.:	6.12

珠海横琴华发世纪广场
Huafa Century Plaza, Hengqin, Zhuhai

地点：中国，珠海
业主：珠海十字门中央商务区建设控股有限公司
设计范围：方案设计
设计时间：2018

Location: Zhuhai, China
Client: Zhuhai Shizimen CBD Construction Holdings Co., Ltd.
Design scope: Proposal Design
Design time: 2018

总平面图 Site Plan

华发世纪广场定位为高端高层住宅项目，位于横琴十字门中央商务区环岛东路与琴海东路之间的十字门大道西侧。基地地处水系入海口与城市主要道路交叉口，并毗邻海岸，同时与澳门隔海相望。

华发世纪广场拥有15座塔楼，户型包括从75 m²到200 m²的多种户型。沿街立面的折线造型将三个地块充分结合在一起，形成一个整体。同时，全部户型都拥有良好的景观朝向和通风。

设计考虑到每个房间的角度，使得每个房间都有良好的视野。每栋楼的每一户都能看到海景。地面部分设计有联排别墅。屋顶花园错落有致，为住户提供公共的绿色活动空间。

设计通过交错穿插的建筑体块形成数个架空空间，两者的对比营造出丰富的建筑立面。同时，体块间的空隙和架空空间将景观引入内部，形成景观通廊和风廊，并可作为公共空间和屋顶花园，提升了整体居住环境的品质。

Huafa Century Plaza is positioned as a high-quality and high rise residential project. It is located on the west side of Shizimen Avenue between Huandao East Road and Qinhai East Road in Shizimen CBD of Hengqin. The base is located at the intersection of the estuary and the main road of the city. It is adjacent to the coast, and is situated across the sea from Macao.

Huafa Century Plaza has a total of 15 towers, and the unit types include a variety of units ranging from 75 m² to 200 m². The polyline shape along the facade of the street fully combines the three plots to form a whole, and at the same time, all the units have good landscape orientations and ventilations.

The design takes the angle of each room into account, so that each room has a good view. Every household has a view of the sea, and the ground section is designed as townhouses. The roof gardens are scattered and provide a public green activity space for the inhabitants.

The design uses inter-weaved building blocks to form several overhead spaces, and the contrast between them creates a rich building facade. At the same time, the voids and overhead spaces between the blocks introduce the landscape inward, forming landscape corridors and wind corridors. These spaces can be used as public spaces and roof gardens to improve the quality of the living environment.

总用地面积：	9.1 hm²
总建筑面积：	220 000 m²
容积率：	2.44

Site area:	9.1 hm²
Gross floor area:	220,000 m²
F.A.R.:	2.44

珠海横琴华发世界汇35#地块
Lot 35# of Huafa World Centre, Hengqin, Zhuhai

地点：中国，珠海
业主：珠海华发房地产开发有限公司
设计范围：方案设计
设计时间：2018

Location: Zhuhai, China
Client: Zhuhai Huafa Real Estate Development Co., Ltd.
Design scope: Proposal Design
Design time: 2018

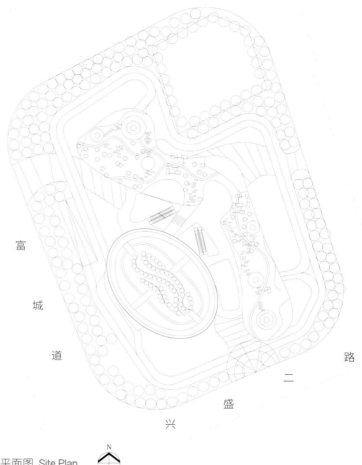

富城道路二盛兴

总平面图 Site Plan

本项目位于横琴十字门中央商务区南区。项目致力于创造场地与东侧和北侧住宅地块、南侧华发世界汇以及西侧商业办公地块相容的连续肌理。

设计受到珠海当地的滨水文化、渔业文化以及高新技术的启发，抽象地提取设计元素，建立一体化的生活模式。"足不出户，服务全世界"这一主题将整个城市联系在一起。

裙房的造型受到当地渔业文化的启发，营造出特别的商业氛围与社区空间。设计与城市的露天活动交融在一起，将公共空间转变成了社区的户外客厅。

高层塔楼同样受到当地滨水文化和渔业文化的启发，其造型像"鱼篓"一样。塔楼高度的变化共同创造出动感的天际线。这些塔楼使底层的社区空间充满活力。设计还通过绿色建筑技术打造生态建筑，使得整个项目成为健康生活的新标志。

The project is located on the south of Shizimen CBD, Hengqin. The project is committed to creating a continuous texture that is compatible with the residential plots on the east and north, Huafa World Plaza on the south and commercial office plot on the west.

The design is inspired by the local aqua culture, fishing culture and advanced technology of Zhuhai. Abstract design elements are extracted, and an integrated life model is established. The theme "serve the world without travel" unites the whole city.

The style of the podium is inspired by local fishing culture. It creates special commercial atmosphere and community space. By merging outdoor activities, the design changes the public spaces into an "outdoor living room" for the community.

The high-rise towers are also inspired by the local aqua and fishing culture. The shapes of these towers have been designed to resemble the "fishing basket". The different heights of the towers work together to create a dynamic skyline. The towers give the ground level community space a lively atmosphere. The design also uses green building techniques to create ecological buildings and makes the whole project a new symbol of healthy life.

总用地面积：	1.4 hm²
总建筑面积：	72 000 m²
容积率：	5.17

Site area:	1.4 hm²
Gross floor area:	72,000 m²
F.A.R.:	5.17

一层平面图 1st Floor Plan

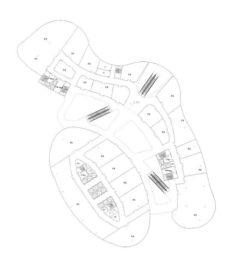

二层平面图 2nd Floor Plan

三层平面图 3rd Floor Plan

四层平面图 4th Floor Plan

上海浦东加拿大梦加园
Dream Home Canada, Pudong, Shanghai

地点：中国，上海
业主：加拿大林业创新投资公司
设计范围：方案设计、扩初设计
设计时间：2003
建成时间：2006

Location: Shanghai, China
Client: Canada Forestry Innovation Investment Company
Design scope: Proposal Design, Design Development
Design time: 2003
Completion time: 2006

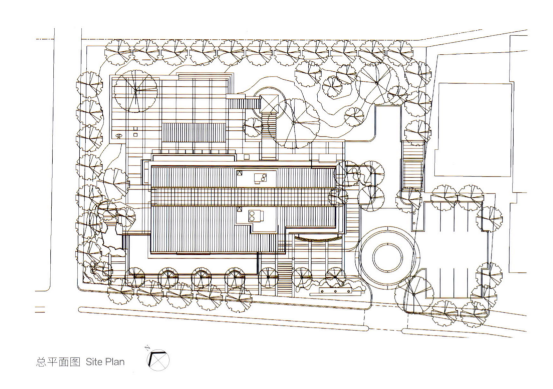

总平面图 Site Plan

梦加园作为一个示范工程，除了传统的产品展厅形式外，其本身即展示了木结构的产品和技术在别墅式住宅和多户组合住宅中的运用，包括内、外部的木装修材料以及非住宅产品。

本项目的主体是894 m²的多功能展示中心。该中心强调了木材产品的技术、结构的高级运用，以及不列颠哥伦比亚省林业项目的展示空间。

As a demonstration project, in addition to the traditional form of product exhibition hall, the Dream Home Canada shows the application of wood structure products and technologies in villa-style houses and multi-family houses, including internal and external wood decoration materials and non-residential products.

The main body of this project is a multifunctional exhibition centre of 894 m². The centre emphasizes the advanced use of wood product technology and structure, as well as a display space for the British Columbia forestry projects.

Site area:	0.3 hm²
Gross floor area:	1,500 m²
Building density:	15%

总用地面积:	0.3 hm²
总建筑面积:	1 500 m²
建筑密度:	15%

上海宝山海纳科技研发大楼
Haina Hi-tech Building, Baoshan, Shanghai

地点：中国，上海
业主：上海大华集团有限公司
设计范围：方案设计、扩初设计
设计时间：2002
建成时间：2005

Location: Shanghai, China
Client: Shanghai Dahua Group Co., Ltd.
Design scope: Proposal Design, Design Development
Design time: 2002
Completion time: 2005

地点：中国，上海
业主：上海大华集团有限公司

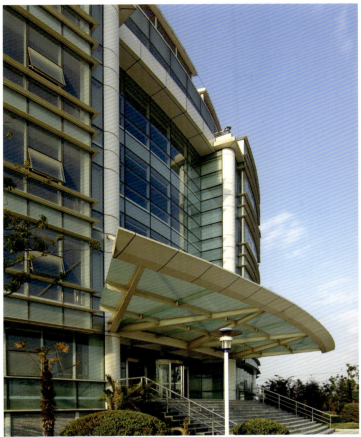

上海大学科技园四通纳米港是为纳米科技提供的研发基地。作为四通纳米港中的标志性建筑，业主希望该建筑体能够成为科技进步与社会发展的一个象征。

6层高的建筑围合一个中庭和一个贝壳形的会展商务中心。开敞的中庭、对称的布局、围合的空间，设计用这样的建筑形式来体现"海纳百川"的意味，确定了办公楼的基本建筑形态。

建筑中央布置有一个大型的多功能空间，它本身的形象就像一个孕育着明珠的贝壳，暗喻纳米港对科技的孵化作用。环形的办公部分围合着一个敞开的中庭空间，支撑空中走廊的两大立柱也由此形成具有科技意味的入口形式，成为纳米港的一个象征性的门户，也使建筑的整体外观更为壮观。南广场的地坪标高做了一个提升，使主要入口更为引人注目。在有限的基地范围内，更多的绿地及休闲空间被设计了出来。

Shanghai University Science Park of Sitong Nanometre Harbor is a research and development base for nanotechnology. As a landmark building in Sitong Nanometre Harbor, the owner hopes that the building can become a symbol of technological progress and social development.

The 6-storey architecture encloses an atrium and a shell-shaped exhibition and business centre. With an open atrium, a symmetrical layout, and an enclosed space, the design uses this building type to reflect the meaning of "inclusive of all rivers", and determines the basic architectural form of the office building.

In the centre of the building, there is a large-scale multi-functional space. Its image is like a shell conceiving a pearl, which is a metaphor for the incubation of the Nanometre Harbor to technology. The circular office part encloses an open atrium space and the two pillars supporting the sky corridor form a technologically meaningful entrance, becoming a symbolic gateway to Nanometre Harbor, making the overall appearance of the building more spectacular. The floor elevation of the south square has been raised to make the main entrance more eye-catching. Within the limited site area, more green spaces and leisure spaces are designed.

总用地面积：	1.1 hm²
总建筑面积：	8 000 m²
容积率：	0.7

Site area:	1.1 hm²
Gross floor area:	8,000 m²
F.A.R.:	0.7

上海静安加拿大KFS国际建筑师事务所办公楼
Office of KFS Architects International Inc. Canada, Jing'an, Shanghai

地点：中国，上海
业主：加拿大KFS国际建筑师事务所
设计范围：方案设计、扩初设计
设计时间：2005
建成时间：2008

Location: Shanghai, China
Client: KFS Architects International Inc.
Design scope: Proposal Design, Design Development
Design time: 2005
Completion time: 2008

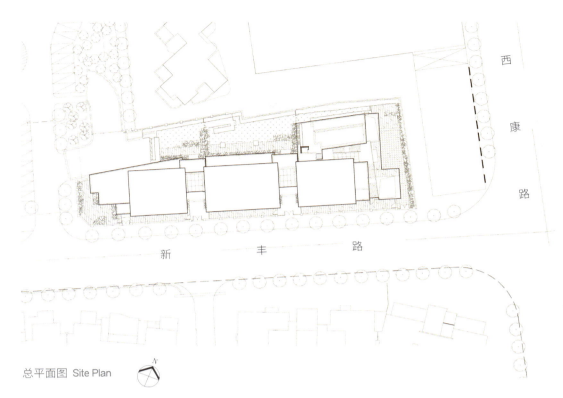

总平面图 Site Plan

该建筑包含三个独立的办公单元,除拥有各自的出入口、沿街面和地下车库外,三个单元之间还通过中庭相连接。这为最终的功能布局带来极大的灵活性。三层的办公单元周边设有精致的景观园林。立面形式采用红砖外墙、现代的玻璃幕墙和局部退台相结合的新古典主义手法,创造出一种全新的小型办公设计模式。

The building is composed of three individual office units. Besides separate independent entrances and exits, individual facades as well as individual underground parking, the three units are connected by atriums. This offers great flexibility for the final functional layout. The 3-storey office units are accompanied with delicate landscapes. The facade is designed in a Neoclassical style comprised of red brick masonry, modern curtain wall and different level terraces, to create a new model of small office design.

| 总用地面积: | 0.2 hm² |
| 总建筑面积: | 3 000 m² |

| Site area: | 0.2 hm² |
| Gross floor area: | 3,000 m² |

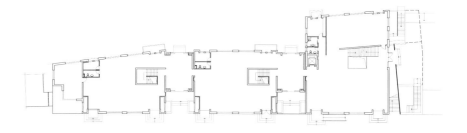

底层平面图 1st Floor Plan

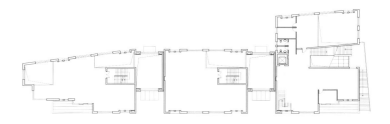

二层平面图 2nd Floor Plan

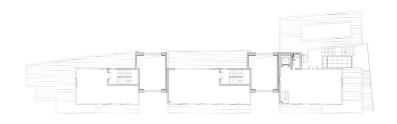

三层平面图 3rd Floor Plan

立面图 Elevation

珠海香洲派诺科技园
Pilot High-tech Park, Xiangzhou, Zhuhai

地点：中国，珠海
业主：珠海派诺科技股份有限公司
设计范围：方案设计、扩初设计
设计时间：2014
建成时间：2019

Location: Zhuhai, China
Client: Zhuhai Pilot Technology Co., Ltd.
Design scope: Proposal Design, Design Development
Design time: 2014
Completion time: 2019

基地位于珠海市香洲区科技创新海岸南围片区。基地内地势平坦，形状为矩形，周围都为已建成的工业用房。

设计坚持三大原则：以节俭为设计策略；以常识为设计基点；以适宜技术为设计手段。

设计从理性的场地分析、建筑体量和空间融合出发。基地主出入口位于西侧。两栋主楼之间为景观水池。设计的地面一层部分为架空层，屋顶为层层跌落的花园。设计运用绿色节能的设计理念。建筑根据风向来设计，以使风的影响最小化。

场地中央设置有两个大型景观水池，与入口广场、步道和首层架空区域相结合，提供了优美的景观。

The site is located in Nanwei Area, Science and Technology Innovation Coast, Xiangzhou District, Zhuhai. The site terrain is flat, with rectangular shape and surrounded by industrial buildings.

The design follows three principles: design strategy with economy, design basis with common sense and design measure with adaptive technology.

The design starts from rational analysis of the site, building scale and space interaction. The main entrance is located in the west. A landscape pond is located between the two main buildings. The design has an elevated ground floor and a stepped roof garden. It incorporates green energy efficiency. The buildings are designed according to the wind direction so as to minimize wind impact.

Two large scale landscape ponds are arranged in the centre of the site linked with the entrance plaza, pedestrian street and elevated ground level, providing a joyful landscape.

总用地面积：	1.2 hm²
总建筑面积：	21 600 m²
容积率：	1.8

Site area:	1.2 hm²
Goss floor area:	21,600 m²
F.A.R.:	1.8

上海静安苏河一号华森钻石广场
1# Suzhou Creek Huasen Diamond Plaza, Jing'an, Shanghai

地点：中国，上海
业主：上海丹林房地产开发有限公司
设计范围：方案设计、扩初设计
设计时间：2005
建成时间：2010

Location: Shanghai, China
Client: Shanghai Danlin Real Estate Co., Ltd.
Design scope: Proposal Design, Design Development
Design time: 2005
Completion time: 2010

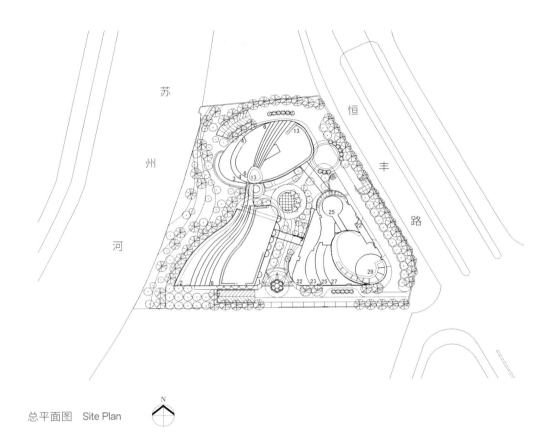

总平面图　Site Plan

N

苏州河

恒丰路

1# Suzhou Creek Huasen Diamond Plaza is in the centre of Shanghai. From the effective use of the internal space of the building to the display of the Suzhou Creek environment and architectural art, the design strives to create a comprehensive and multi-functional community to achieve the coexistence of the environment and the architecture.

苏河一号华森钻石广场位于上海市中心区内。本设计从建筑内部空间的有效利用到苏州河环境与建筑艺术的展示，力求创造一个全方位、多功能的社区，以达到环境与建筑的融合共存。

沿苏州河边有45°的高度限制，因此设计将三幢楼由低至高、由西向东呈阶梯状布置。其中两幢沿河在南北方向上展开，以获得最大的景观展开面。流线型的建筑体量层层跌落，形似天空中的云朵，减轻了高层建筑对河道及开放空间的压抑感。

东侧办公楼和西侧公寓式酒店及办公综合体之间布置有一条商业街，具有交通走廊、消防通道及休闲空间的功能。中庭及天桥连廊将建筑连成一线，为不同区域提供了不同的空间尺度。商业、办公、休闲、娱乐、景观形成了一个24小时国际商务社区。

The design of the three buildings is a step-up from west to east following the 45° height control adjacent to the Suzhou Creek. Two buildings are arranged from north to south for the maximum landscape exposure to the Suzhou Creek. Streamlined built volumes fall fluidly on one after another like clouds in the sky, mitigating the oppressive feeling of the high-rise buildings to the creek and open space.

A commercial street, between the office building on the east and the apartment hotel and office complex on the west, serves as the transportation corridor, fire engine access and recreational space. Atriums and a pedestrian bridge connect the buildings and offer a variety of spatial scales to different areas. The retail, office, leisure, recreation and landscape generate an 24-hour international business community.

总用地面积：	1.4 hm²
总建筑面积：	68 000 m²
容积率：	5.0

Site area:	1.4 hm²
Gross floor area:	68,000 m²
F.A.R.:	5.0

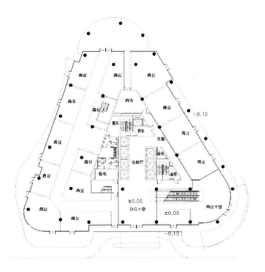

一层平面图 1st Floor Plan

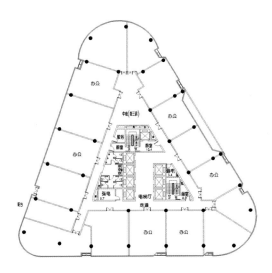

二层平面图 2nd Floor Plan

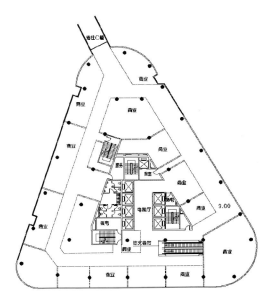

办公楼典型平面 Office Tower Typical Floor Plan

上海静安达安河畔雅苑
Da'an Riverside Tower, Jing'an, Shanghai

地点：中国，上海
业主：上海达安房产开发有限公司
设计范围：方案设计、扩初设计
设计时间：2003
建成时间：2006

Location: Shanghai, China
Client: Shanghai Da'an Real Estate Development Co., Ltd.
Design scope: Proposal Design, Design Development
Design time: 2003
Completion time: 2006

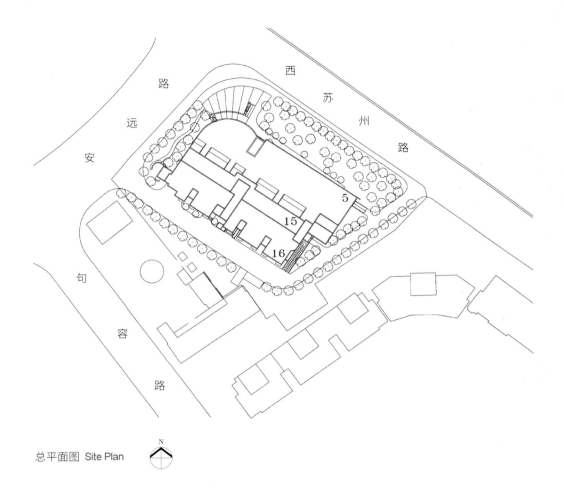

总平面图 Site Plan

本项目位于上海市静安区苏州河边。项目意在创建一个高品质的小户型办公、商业与住宅综合体，建设一个环境优美、设施完备、建筑风格独特的国际化社区。

简约的总体布局形成了富有戏剧性的空间，使得景色变化多样。苏州河上让出的大量空间，为基地留下更多的阳光，也为城市主干道留下更多的可用空间。

70~90 m²的公寓采用跃层式布局，让每户居民既有南向日照，又能欣赏到北侧的苏州河景观。设计的灵活性使部分单元可以组合成面积更大的房型。层层退台的建筑体量，制造了多个顶层露台和顶层公寓，丰富了建筑物的建筑语汇。

This project stands along the Suzhou Creek in Jing'an District, Shanghai. It creates a small-scale complex of offices, commercial spaces and residences of high quality. This complex forms an international community with a graceful environment, complete facilities and a distinctive architectural style.

A simplified master plan forms a dramatic space with a variety of changing views. Large space is set back from Suzhou Creek allowing more sunlight into the site and providing the city street with more usable space.

The 70~90 m² apartments are arranged in a duplex system which offers sunlight exposure from the south and Suzhou Creek views from the north to the residents. Design flexibility makes it possible for some apartments to be incorporated into larger units. The "step design" of the building volume creates many roof terraces and penthouses that enrich the architectural vocabulary of the building.

总用地面积：	0.43 hm²
总建筑面积：	16 000 m²
容积率：	3.1

Site area:	0.43 hm²
Gross floor area:	16,000 m²
F.A.R.:	3.1

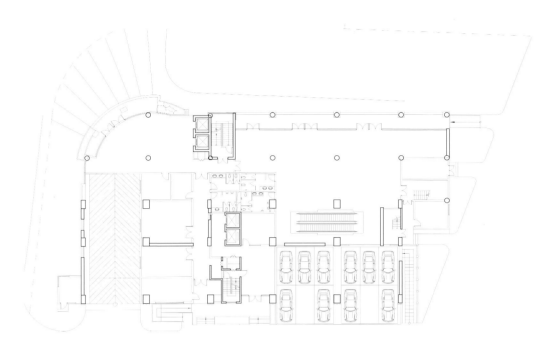

底层平面图 1st Floor Plan

标准层平面图 Typical Floor Plan

珠海香洲金山软件园
Kingsoft Headquarters, Xiangzhou, Zhuhai

地点：中国，珠海
业主：金山软件股份有限公司
设计范围：方案设计、扩初设计
设计时间：2009
建成时间：2019

Location: Zhuhai, China
Client: Kingsoft Corporation Limited
Design scope: Proposal Design, Design Development
Design time: 2009
Completion time: 2019

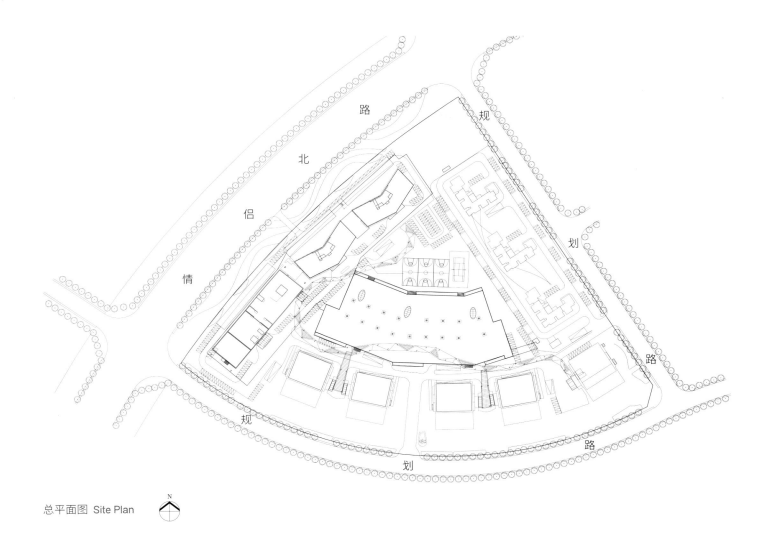

总平面图 Site Plan

N

金山软件园位于珠海高新区总部基地西南侧的唐家湾地块。基地南邻大海，东北依石坑山，景观环境优越。

建筑设计的灵感来自群山和海浪。

1. 成熟的技术
时尚的建筑形式和简洁的结构布局方式营造出最有效的功能空间，有利于节约成本、分期建设以及缩短建设周期。

2. 大片的中心绿化景观
建筑沿周边布局，形成大片的中央绿化景观用作公共空间。

3. 超级海景
主要建筑物沿路规划，以保证最大的朝海面。建筑设置似双手环抱着基地，最大限度利用天然的观景面。

总用地面积：	9.7 hm²
总建筑面积：	145 000 m²
容积率：	1.5

Kingsoft Headquarters is located in Tangjiawan plot, southwest of Zhuhai High-Tech Zone Headquarters Base. The site is adjacent to the sea in the south and Shikeng Mountain in the northeast. It has a spectacular landscape environment.

The architectural design is inspired by the surrounding mountains and the waves.

1. Mature technology
The most functional spaces are created by fashionable architectural form and simple structural layout, which are beneficial to cost saving, phased construction and shortened construction cycles.

2. Large central green landscape
The building is arranged along the periphery to form a large central green landscape as a public space.

3. Super ocean view
The main buildings are planned along the road to ensure the maximum sea frontage. The setting of the buildings is like embracing the site with arms to maximize the use of the natural viewing surface.

Site area:	9.7 hm²
Gross floor area:	145,000 m²
F.A.R.:	1.5

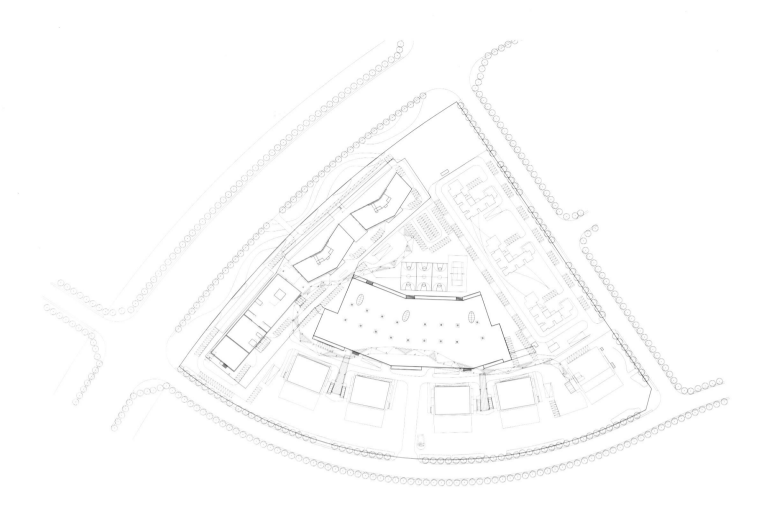

一层平面图 1st Floor Plan

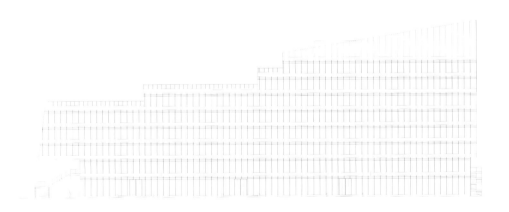

4# — 8# 立面图 Elevation of 4# — 8#

杭州西湖勾山里
Goushan Li, Xihu, Hangzhou

地点：中国，杭州
业主：杭州涌金置业投资有限公司
设计范围：方案设计、扩初设计、施工图设计
设计时间：2009
建成时间：2015

Location: Hangzhou, China
Client: Hangzhou Yongjin Real Estate Investment Co., Ltd.
Design scope: Proposal Design, Design Development, Construction Documentation
Design time: 2009
Completion time: 2015

总平面图 Site Plan

本项目与著名的西湖景区和中国美术学院等杭州优秀建筑相邻。设计保留了基地内部的四栋老建筑，并满足了建筑的15 m的限高。

总平面图的设计将业主和城市公共利益做了平衡处理。设计使用了表达稳重特点的形体处理。为了协调的周边环境，立面材料也是精心选择的。

This project is in close proximity to the famous tourist attraction of West Lake, China Academy of Art and other outstanding architectures of Hangzhou. The design preserves four historical buildings in the site and meets the 15 m building height limitation.

The site plan conveys a balance of the private owner and the city public's interests. The design uses a design language that expresses characteristics of stability through shape and form. For a harmonious surrounding environment the facade materials are chosen with great care.

总用地面积:	1.3 hm²
地上建筑面积:	15 100 m²
地下建筑面积:	16 200 m²
容积率:	1.1

Site area:	1.3 hm²
Above ground G.F.A.:	15,100 m²
Underground G.F.A.:	16,200 m²
F.A.R.:	1.1

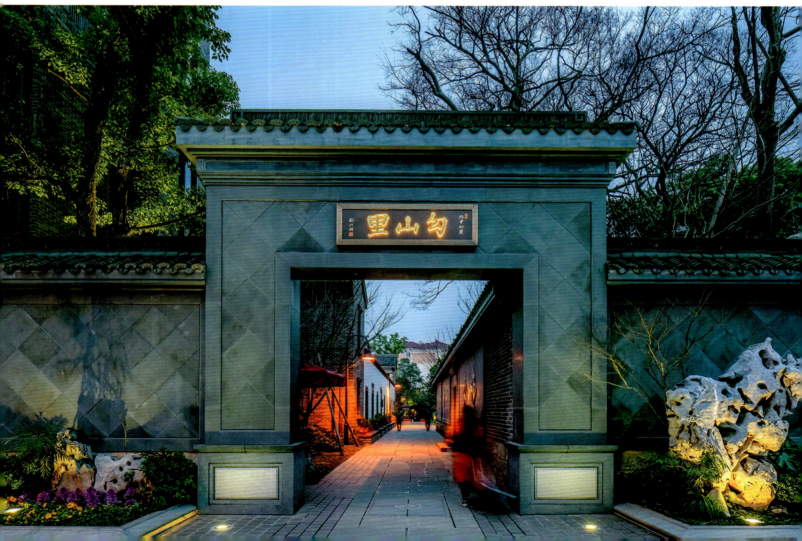

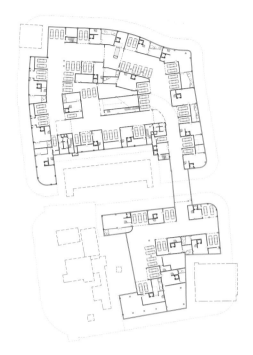

地下二层平面图 B2 Floor Plan

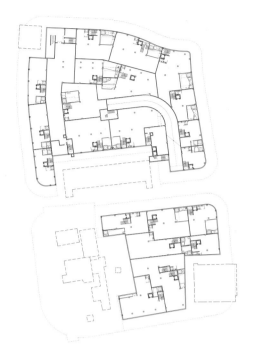

地下一层平面图 B1 Floor Plan

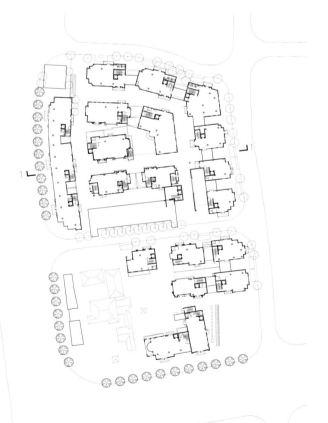

一层平面图 1st Floor Plan

二层平面图 2nd Floor Plan

苏州吴中唯亭君地新大陆广场
Weiting Jundi New Continental Plaza, Wuzhong, Suzhou

地点：中国，苏州
业主：上海君地投资有限公司
设计范围：方案设计、扩初设计
设计时间：2010
建成时间：2014

Location: Suzhou, China
Client: Shanghai Jundi Investment Co., Ltd.
Design scope: Proposal Design, Design Development
Design time: 2010
Completion time: 2014

唯 华 路

亭 隆 街

阳 澄 湖 大 街

总平面图 Site Plan

本项目采用"化零为整"的设计理念，将项目建筑沿街道布置，形成整齐划一的城市元素。

六栋商业服务塔楼位于绿地南北两侧。一层的商铺在街面连成一个整体。中央绿地成为主要的景观特色，以乔木和大片草地为主，配以广场，形成丰富多变的休闲空间。商业部分以硬质铺装为主，配合休闲椅，营造出舒适的空间。

立面采用装饰艺术风格，强调竖向线条和顶部收分，让60 m高的建筑显得更加挺拔。设计十分注重立面细部材料的选择，体现出高贵、典雅的品质。

"Gather parts into a whole" is used as the design concept of this project, and the structures of this project are placed along the street, creating an orderly and unified urban element.

Six commercial service towers are positioned on the north and south sides of the green space. The one-storey shops are connected as a whole along the street. The central green space acts as the major landscaping feature mainly consisting of large trees and fields of grass. Equipped with a square, it forms a rich and varied leisure space. The commercial part is covered mainly by hard pavements, with leisure chairs to create a comfortable space.

The facades are inspired by the art-deco style, emphasizing vertical lines with volumetric shrinkage at the top, giving the 60 m building a greater sense of height. The facade detailing is carefully considered through the choice of materials, displaying the qualities of luxury and elegance.

总用地面积：	3.7 hm²	Site area:	3.7 hm²
总建筑面积：	149 000 m²	Gross floor area:	149,000 m²
容积率：	3.03	F.A.R.:	3.03

苏州吴中唯亭君地曼哈顿广场
Weiting Jundi Manhattan Plaza, Wuzhong, Suzhou

地点：中国，苏州
业主：上海君地投资有限公司
设计范围：方案设计、扩初设计
设计时间：2013
建成时间：2016

Location: Suzhou, China
Client: Shanghai Jundi Investment Co., Ltd.
Design scope: Proposal Design, Design Development
Design time: 2013
Completion time: 2016

唯 华 路
阳 澄 湖 大 道

总平面图 Site Plan

基地南邻阳澄湖大道，北邻唯华路，西邻亭盛街，东邻亭隆街。基地东侧为已建成的君地新大陆广场。阳澄湖大道是一条交通主干道，交通十分便利。

基地所在区域处于青剑湖和阳澄湖之间，北侧沿线视野在一定高度和角度有较好景观。

设计充分尊重城市肌理的现状。君地新大陆广场是一个主轴对称的项目，这根轴线连接青剑湖商业中心。基地西侧为轴线上的唯亭医院。轴线沿阳澄湖大道，在青剑湖与阳澄湖之间形成了一个整体效果。因此，本项目的三栋塔楼也对称布局在这根轴线上，力求完善阳澄湖大道的城市景观。

The site is on the north of Yangcheng Lake Avenue, south of Weihua Road, east of Tingsheng Street, and west of Tinglong Street. On the east of the site is the built Jundi New Continental Plaza. Yangcheng Lake Avenue is a main road that makes the traffic convenient.

The site is located between Qingjian Lake and Yangcheng Lake, with good views of landscapes along the north side at a certain height and angle.

The design fully respects the status quo of the urban texture. Jundi New Continental Plaza is an axially symmetrical project where the axis points to the commercial centre of Qingjian Lake. West of the site is Weiting Hospital on the axis. The axis forms a whole expression along Yangcheng Lake Avenue in between Qingjian Lake and Yangcheng Lake. Therefore, the three towers of this project are also arranged symmetrically on the axis to make the urban landscape of Yangcheng Lake Avenue complete.

总用地面积：	3.6 hm²
总建筑面积：	90 000 m²
容积率：	2.5

Site area:	3.6 hm²
Gross floor area:	90,000 m²
F.A.R.:	2.5

张家港金港君地B52#地块
Lot B52# Jundi, Jin'gang, Zhangjiagang

地点：中国，张家港
业主：张家港君地房地产开发有限公司
设计范围：方案设计、扩初设计
设计时间：2020

Location: Zhangjiagang, China
Client: Zhangjiagang Jundi Real Estate Development Co., Ltd.
Design scope: Proposal Design, Design Development
Design time: 2020

B52#地块位于张家港金港路西侧，晨丰公路北侧。基地为三角形地块，用地性质为商业用地。

建筑两边衔接城市干道，设计了多功能的临街商业门面，倡导一种既时尚又个性的生活方式，并体现出鲜明的时代特征。

立面风格参考纽约曼哈顿时代广场的建筑风格，简化雕琢和装饰，强调线条感。在设计中，现代建筑语汇经提炼后加以运用，并和现代建筑材料结合在一起，配以明快的色彩，形成富有韵律感的立面和空间效果。

Lot B52# is located at the west of Jin'gang Road and north of Chenfeng Road, Zhangjiagang. The site presents in a triangular shape and the land is planned for commercial use.

The two sides of the building are connected to city streets, designed with multi-functional stores facing the street that propose a fashionable and characteristic lifestyle, reflecting a distinctive characteristic of the times.

The facade style refers to the style of Time Square in Manhattan, New York. The decorations and ornaments are simplified to emphasize a linear expression. Within the design, modern architectural language is distilled and applied in a combination of modern materials and bright colours to form a rhythmic facade and spacial effect.

总用地面积：	1.1 hm²
总建筑面积：	19 364.35 m²
容积率：	1.20

Site area:	1.1 hm²
Gross floor area:	19,364.35 m²
F.A.R.:	1.20

龙口滨海旅游度假区
Coastal Tourist Resort, Longkou

地点：中国，龙口
业主：烟台佳安房地产开发有限公司
设计范围：方案设计
设计时间：2010

Location: Longkou, China
Client: Yantai Jia'an Real Estate Development Co., Ltd.
Design scope: Proposal Design
Design time: 2010

本项目为龙口市滨海旅游度假区的17#地块，用地地势平坦，距北侧渤海海岸仅500 m。

基地景观资源极其优越，东侧和北侧被松树林景区所围绕。

从景观价值最大化的角度考虑，高层建筑沿基地北侧布置，并高低错落，形成滨海沿线连续起伏的天际线。

多层度假公寓沿基地南侧布置，位于高层建筑与低层会所之间。一连串建筑高度的变化丰富了空间层次。

The project is located on Lot 17# in Coastal Tourist Resort, Longkou. The site is flat and is only 500 m to the coastline of the Bohai Sea.

The site possesses an abundance of landscape resources and is surrounded by pine forests scenic area to the east and north.

From the perspective of maximizing the value of the landscape, high-rise buildings are arranged along the north side of the site, with differences in height to create a continuous and undulant coastal skyline.

Multi-storey resort apartments are arranged along the south side of the site between the high-rise buildings and the low-rise club. A series of uprising volumes enrich the spatial level.

总用地面积:	9.74 hm²
总建筑面积:	136 398 m²
容积率:	1.4

Site area:	9.74 hm²
Gross floor area:	136,398 m²
F.A.R.:	1.4

上海青浦港隆广场
Glory Harbour, Qingpu, Shanghai

地点：中国，上海
业主：上海风翔房地产开发有限公司
设计范围：方案设计、扩初设计
设计时间：2005
建成时间：2008

Location: Shanghai, China
Client: Shanghai Fengxiang Real Estate Development Co., Ltd.
Design scope: Proposal Design, Design Development
Design time: 2005
Completion time: 2008

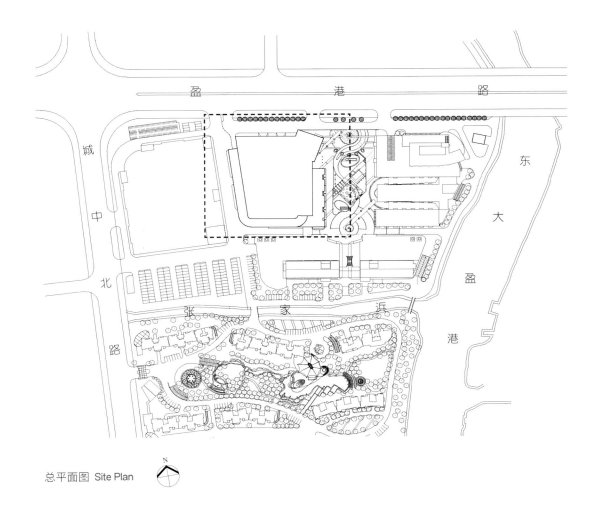

总平面图 Site Plan

本项目位于上海青浦，靠近盈港路与城中北路，南侧为张家浜，东侧为东大盈港。项目在地理位置与交通方面均有优势，且西侧已建有大型超市。两条河流强化了基地的水景优势。基地周边还有商业、办公等多栋建筑。

港隆广场设置了大面积的休闲空间。绿化广场以植物、草坪、水道构成极具现代气息的室外空间。音乐喷泉位于广场正中，是整个商业中心的视觉焦点。设计营造出一个花园式的购物环境，既为城市增光添彩，也塑造了新一代自然环保的公共空间，为改善城市生活做出了一定的贡献。

The project is located at Qingpu, Shanghai and is close to Yinggang Road and Chengzhong North Road. Zhangjia Brook is located at east and Dongdaying Gang River lies at south. The site has both geographic and transportation advantages. There is a built supermarket to the west of the site. The two rivers enhance the waterscape scene advantage to the site. There are also a number of commercial and office blocks near the project area.

The Glory Harbour has a large area for leisure. This green plaza is a modern style outdoor place due to its plants, lawns and water lanes. The musical fountain, which is located at the centre of the plaza, acts as the focus point of the whole commercial area. The design offers a garden-like shopping environment. It boosts the reputation of the city, presents a new generation of eco-green public space and contributes to a better city life.

总用地面积:	8.32 hm²
总建筑面积:	166 116 m²
容积率:	2.0

Site area:	8.32 hm²
Gross floor area:	166,116 m²
F.A.R.:	2.0

珠海横琴麦吉尔大学环球工作站
Global Workstation of McGill University, Hengqin, Zhuhai

地点：中国，珠海
业主：加拿大麦吉尔大学傅国华建筑学院 / 珠海大横琴置业有限公司
设计范围：方案设计
设计时间：2019

Location: Zhuhai, China
Client: Peter Guo-hua Fu School of Architecture McGill University / Zhuhai Dahengqin Real Estate Co., Ltd.
Design scope: Proposal Design
Design time: 2019

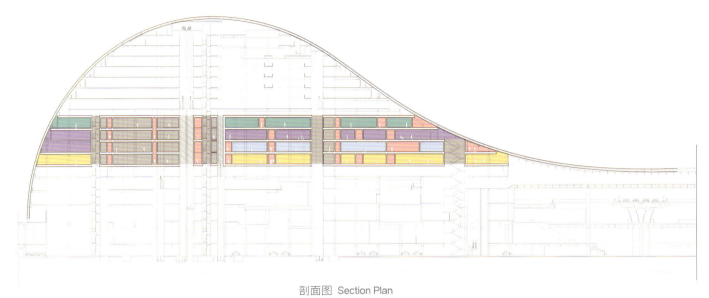

剖面图 Section Plan

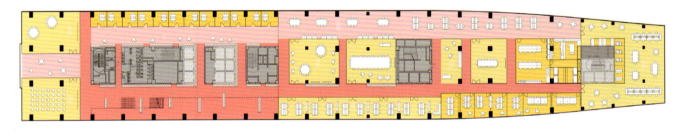

七层平面图 7th Floor Plan

此项目是麦吉尔大学傅国华建筑学院的实习工程项目。

麦吉尔大学环球工作站拟建于横琴"天沐琴台"内，共占据四层空间。主要功能包括共享空间、图书馆、学习空间、教学实验室、主动学习教室和休闲空间。

工作站的设计充分利用横琴的亚热带气候，创造全年舒适的户外空间。屋顶花园、露台和阳台的结合，加强了建筑室内外的联系。

充足的自然光线照亮了实验室和公共空间，减少了对人工照明的依赖，从而降低了电力消耗和能源费用，并减少了对环境的影响。

这些策略结合在一起，创造了一种隐蔽的开放感。这些元素在城市的建筑和公共空间中的应用，形成独特的横琴特色，让城市展现出一种轻松的生活方式。

This project is an intern project of McGill University Peter Guo-hua Fu School of Architecture.

The Global Workstation of McGill University is planned in "Tianmu Qintai", occupying four storeys. Main functions include shared space, libraries, study space, teaching laboratories, active learning classrooms, and leisure space.

The workstation is designed to make the most of the sub-tropical climate in Hengqin by creating outdoor spaces that are comfortable all year round. A strong connection between indoor and outdoor is created by combining roof gardens, terraces and balconies.

The abundant natural sunlight illuminates the laboratories and the public space. Natural light reduces the reliance on artificial lighting resulting in reduced electricity consumption, decreased energy bills and less environmental impact.

These strategies combine to create a sense of sheltered openness. The adaptation of these elements into the architectures and public space of the city can contribute to a unique Hengqin identity which will allow the city to express a relaxed lifestyle.

上海长宁愚园路综合改造
Yuyuan Road Renovation, Changning, Shanghai

地点：中国，上海
业主：加拿大麦吉尔大学傅国华建筑学院/上海申亚投资控股有限公司
设计范围：方案设计
设计时间：2018

Location: Shanghai, China
Client: Peter Guo-hua Fu School of Architecture McGill University / Shanghai Shenya Investment Holding Co., Ltd.
Design scope: Proposal Design
Design time: 2018

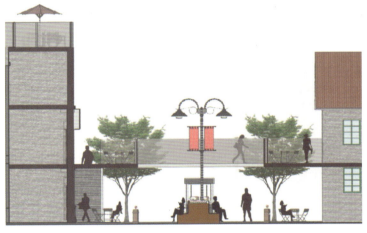

总平面图 Site Plan

N

此项目是麦吉尔大学傅国华建筑学院的实习工程项目，是位于上海市愚园路的改造设计。设计希望通过对基础设施的改善来提升城市空间的品质。

项目基地内有部分名人故居和洋房。愚园路综合改造项目尊重历史，通过对建筑文化遗产的保留，增加文化空间并保留商住互融的传统，建设绿色公共空间等方法来营造具有文化氛围的上海街区。

方案拆除整治了违章搭建，对交通系统进行了梳理优化；新建了文化中心，提升当地的艺术气息；在广场中布置了开放式商业空间、绿化休闲活动区域以及屋顶花园，提升了当地的景观质量。设计打造出将文化传承、创意设计、金融参与、展示空间多方面相结合的创新生态环境。

This project is an intern project of McGill University Peter Guo-hua Fu School of Architecture and a renovation project of Yuyuan Road, Shanghai. The design aims to improve the quality of urban space through the improvement of the infrastructure.

The site contains several residences of former celebrities and Western style houses. The Yuyuan Road renovation project respects the history by preserving architectural cultural heritage, increasing cultural space, preserving the tradition of commercial and residential integration, and building a green public space to create a neighborhood of cultural atmosphere in Shanghai.

The scheme dismantles and rectifies illegal structures and provides a solution that optimizes the traffic system. A new cultural centre is built to enhance the local artistic atmosphere. By arranging open commercial space, green leisure area and roof gardens in the plaza, the quality of local landscape is increased. The design creates an innovative ecological environment combining cultural inheritance, creative design, financial participation and exhibition space.

总用地面积：	5.2 hm²
总建筑面积：	65 260 m²
容积率：	1.25

Site area:	5.2 hm²
Gross floor area:	65,260 m²
F.A.R.:	1.25

RESIDENTIAL
DESIGN

居住建筑

三亚吉阳亚龙湾壹号居住综合体
Residential Complex of 1# Yalong Bay, Jiyang, Sanya

地点：中国，三亚
业主：海南申亚置业有限公司
设计范围：方案设计、扩初设计
设计时间：2013
建成时间：2017

Location: Sanya, China
Client: Hainan Shenya Real Estate Co., Ltd.
Design scope: Proposal Design, Design Development
Design time: 2013
Completion time: 2017

总平面图 Site Plan

亚龙湾壹号居住综合体，即AC-3地块呈扇形，且在扇形中部有一个规划中的景观湖。地形整体自北向南呈缓坡状。地块西南侧被椰风路所环绕，西北侧为拟建设的五星级酒店，北侧为城市市政设施用地，东侧及南侧为规划中的运动休闲场地。

由于AC-3地块远离城市道路，适宜营造静谧的旅游度假区。景观设计营造出带有内部水系的舒适生活环境，或汇入内湖，或流入周边的溪流。临水而居的生活方式将社区完美地与自然环境融为一体。

The Residential Complex of 1# Yalong Bay, also archived as Lot AC-3 has a radial shaped site, with a landscape lake planned in the middle. The overall site slope from north to south. The site is surrounded by Yefeng Road on the southwest, a five-star hotel proposed on the northwest, an urban municipal facility on the north, and a proposed sports recreational area on the east and south.

The location of the Lot AC-3 which is away from city traffic is ideal for the proposal of a peaceful holiday resort. The landscape design aims to create a pleasant living environment with the internal water system which connects either to the inner lake or to the surrounding streams. The community is perfectly integrated with the natural environment with a "living by the water" lifestyle.

总用地面积:	16.3 hm²
总建筑面积:	132 000 m²
容积率:	0.8

Site area:	16.3 hm²
Gross floor area:	132,000 m²
F.A.R.:	0.8

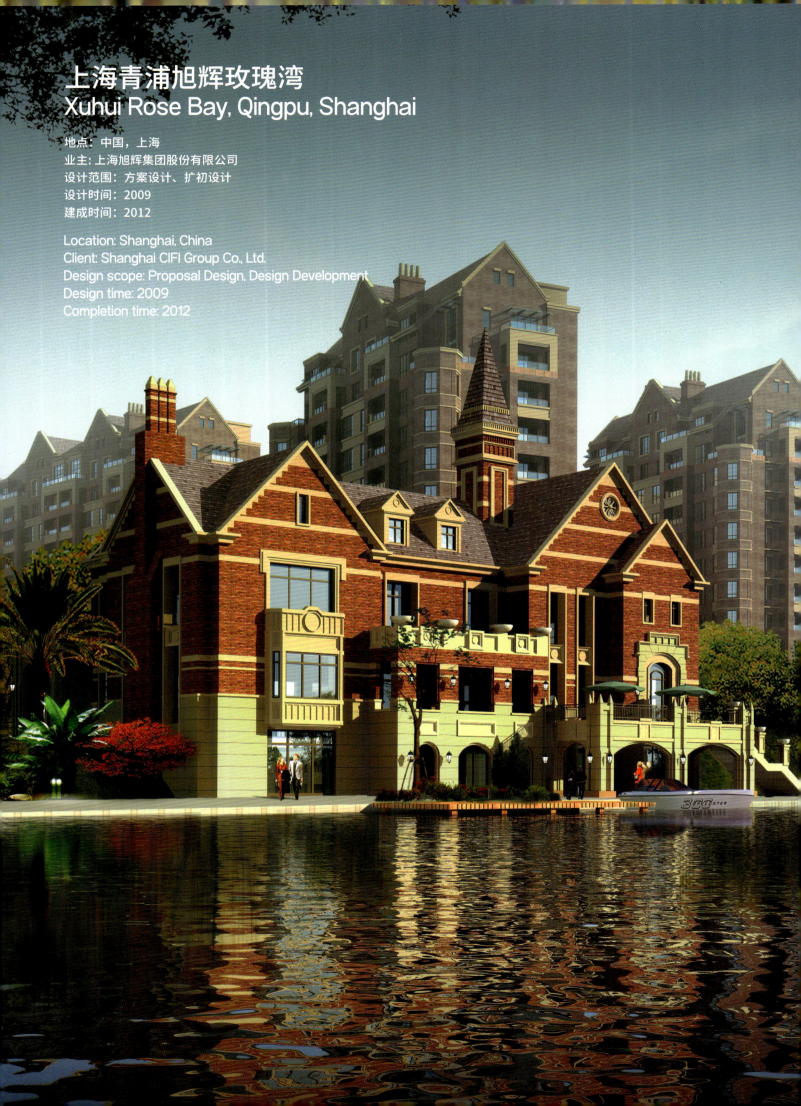

上海青浦旭辉玫瑰湾
Xuhui Rose Bay, Qingpu, Shanghai

地点：中国，上海
业主: 上海旭辉集团股份有限公司
设计范围：方案设计、扩初设计
设计时间：2009
建成时间：2012

Location: Shanghai, China
Client: Shanghai CIFI Group Co., Ltd.
Design scope: Proposal Design, Design Development
Design time: 2009
Completion time: 2012

保亭七仙岭
Qixian Range, Baoting

地点：中国，保亭
业主：北京荣京投资集团有限公司
设计范围：方案设计
设计时间：2011

Location: Baoting, China
Client: Beijing Rongjing Investment Group Co., Ltd.
Design scope: Proposal Design
Design time: 2011

总平面图 Site Plan

本项目距离三亚仅25分钟车程。项目定位为集温泉、休闲、商务、度假为一体的五星级温泉度假区。

本项目以传统地域文化为根基，融入西方文化。亚洲元素被植入现代建筑语系，将传统意境和现代元素对称运用来表现和反映当地的传统。项目在关注现代生活舒适性的同时，让亚洲传统文化得以传承和发扬。

The project is only 25 minutes drive to Sanya. The project is defined as a five-star resort with hot springs, leisure, business and vacation accommodations.

The project is defined by its traditional regional culture with western culture blended in. Asian elements are put into the modern architectural language where traditional image and modern elements are used symmetrically to express and reflect local traditions. The project inherits and develops Asian traditional culture while focusing on the comfortable modern lifestyle.

总用地面积:	16.3 hm²
总建筑面积:	166 800 m²
容积率:	0.8

Site area:	16.3 hm²
Gross floor area:	166,800 m²
F.A.R.:	0.8

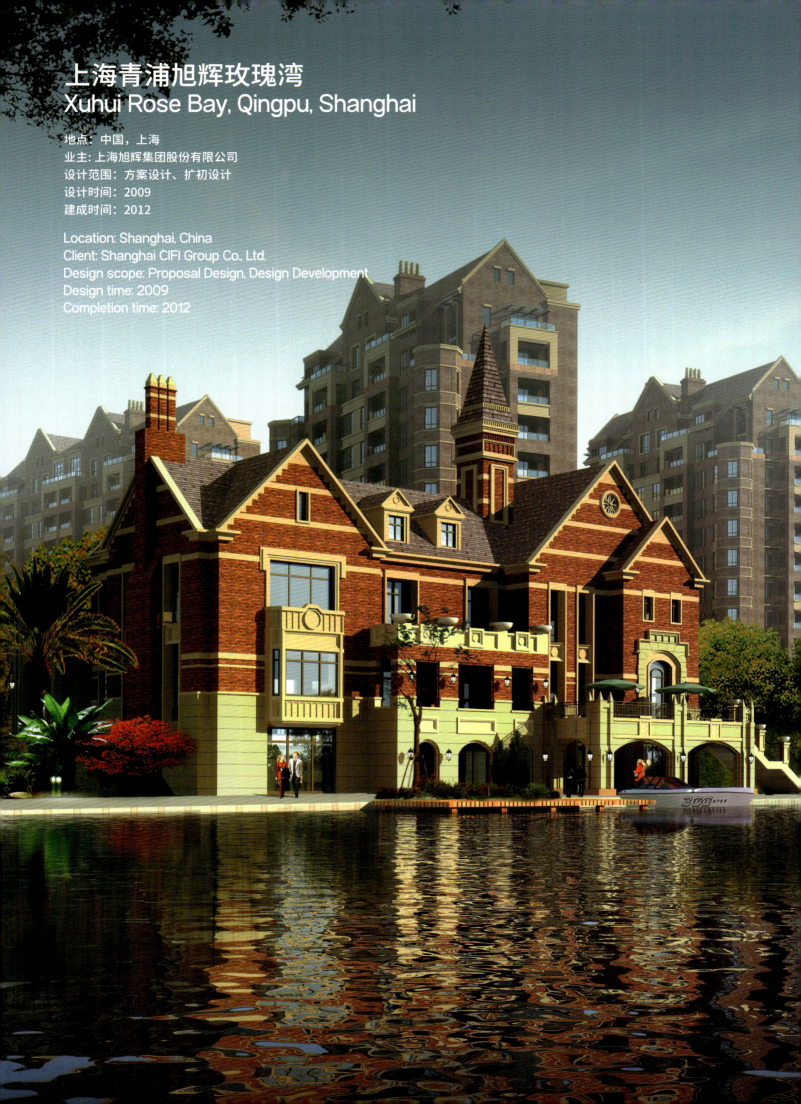

上海青浦旭辉玫瑰湾
Xuhui Rose Bay, Qingpu, Shanghai

地点：中国，上海
业主：上海旭辉集团股份有限公司
设计范围：方案设计、扩初设计
设计时间：2009
建成时间：2012

Location: Shanghai, China
Client: Shanghai CIFI Group Co., Ltd.
Design scope: Proposal Design, Design Development
Design time: 2009
Completion time: 2012

本项目位于青浦工业园区西侧，东至中大盈港，南至清河湾路，西至漕盈路，北至胡渡浜。项目北面临湖，东面环水，生态条件优越。

基地配套设施较为完善，车行10分钟可达老城区商业中心。商业中心内已设有完善的商场、银行、菜市场等商业设施。

基地呈"L"形。水岸线超过600 m，水体清澈，是非常珍贵的自然生态资源。建筑与水体之间的有机对话是此次规划的基本依据。

The project is located on the west side of Qingpu Industrial Park, reaching Zhongdaying Gang River to the east, Qinghewan Road to the south, Caoying Road to the west and Hudu Brook to the north. The project is adjacent to the lake in the north and surrounded by water in the east, where the ecological conditions are desirable.

The supporting facilities of the base are relatively complete. The commercial centre of the old city can be reached by car in 10 minutes. There are complete commercial facilities such as shopping malls, banks and markets in the commercial centre.

The base is in the shape of an "L". The water coastline is more than 600 m long. The clear water is an essential natural ecological resource. The organic dialogue between the architectures and the water is the basis of the planning.

总用地面积:	7.32 hm²
总建筑面积:	163 927 m²
容积率:	1.8

Site area:	7.32 hm²
Gross floor area:	163,927 m²
F.A.R.:	1.8

上海松江旭辉安贝尔花园
Xuhui Albert Garden, Songjiang, Shanghai

地点：中国，上海
业主：上海旭辉集团股份有限公司
设计范围：方案设计、扩初设计
设计时间：2014
建成时间：2017

Location: Shanghai, China
Client: Shanghai CIFI Group Co., Ltd.
Design scope: Proposal Design, Design Development
Design time: 2014
Completion time: 2017

总平面图 Site Plan

基地位于上海松江区车墩镇，紧邻上海影视乐园。地块东至影城路，南至影车路，西至车亭公路，北至影振路。

本项目为中密度居住社区。设计利用周边资源，创造各种不同居住类型的，具有独立个性的居住空间。

总体规划确保每栋建筑都有最佳朝向，同时尽量降低G15高速的噪声影响。设计在北侧与西侧布置多层住宅，东南侧布置低层联排住宅。两者之间通过中央景观带连接。南北向的中央景观轴营造出运动、休闲、交流的空间。沿影振路与影城路布置的商业用房形成了一个30 m宽，50 m长的公共广场，并在影振路上设有主入口，为居民提供了一个开放的空间。

The site is located in Chedun Town of Songjiang District, Shanghai, next to the Shanghai Film Park. The site is bounded by Yingcheng Road on the east, Yingche Road on the south, Cheting Highway on the west, and Yingzhen Road on the north.

The project is a mid-density residential area. The design utilizes the surrounding resources to provide independent and characteristic living space with a wide range of housing typologies.

The overall layout ensures that each building has the best orientation, while at the same time minimizing the noises from the G15 Expressway. The design has multi-storey apartments on the north and west side, and low-rise townhouses on the southeast with a central landscape belt as connection. The north-south central landscape corridor provides spaces for exercise, leisure and communication. The commercial buildings along Yingzhen Road and Yingcheng Road form a 30 m by 50 m public square with a main entrance on the Yingzhen Road, providing open space to the residents.

总用地面积约：	8 hm²
总建筑面积：	129 000 m²
容积率：	1.3

Site area:	8 hm²
Total Building area:	129,000 m²
F.A.R.:	1.3

上海宝山旭辉依云湾
Xuhui La Baie D'Evian, Baoshan, Shanghai

地点：中国，上海
业主：上海旭辉集团股份有限公司
设计范围：方案设计、扩初设计
设计时间：2002
建成时间：2008

Location: Shanghai, China
Client: Shanghai CIFI Group Co., Ltd.
Design scope: Proposal Design, Design Development
Design time: 2002
Completion time: 2008

总平面图 Site Plan

基地位于上海市宝山区顾村地区，北邻顾北路，东邻顾荻路，南面和西面为荻泾河。

设计借助鲜明、古典、华丽的设计理念，以洗练的布局，精致的建筑设计，赏心悦目的景观意象，使此项目从不起眼的地段与环境中脱颖而出，成为宝山地区的一颗耀眼明珠。

住宅产品主要为高层建筑和联排别墅。基地不同方位的特点决定了各种产品的布置。

高层区设置在小区东面，并在顾荻路设有独立出入口，使高层区和别墅区互不干扰。整体空间由西向东升高，保证荻泾河这一主要景观资源能被充分利用。

The site is located in the Gucun area of Baoshan District, Shanghai, with Gubei Road on the north, Gudi Road on the east, and Dijing River on the south and west.

The design draws on distinctive, classical and noble design concepts, through refined layout, exquisite architectural design, and pleasing landscape image, making this project stand out from the humble location and environment and become a dazzling pearl in Baoshan District.

Residential products are mainly high-rises and townhouses. The characteristics of the site in different locations determine the layout of the products.

The high-rise area is set on the east of the residential area with an independent entrance on Gudi Road, so that the high-rise area and the villa area do not interfere with each other. The overall space rises from west to east to ensure that the main landscape resource of Dijing River can be fully utilized.

总用地面积:	18.9 hm²
总建筑面积:	220 000 m²
容积率:	1.2

Site area:	18.9 hm²
Gross floor area:	220,000 m²
F.A.R.:	1.2

Gubei International Garden, Changning, Shanghai

地点：中国，上海
业主：中华企业股份有限公司
设计范围：方案设计、扩初设计
设计时间：2002
建成时间：2006

Location: Shanghai, China
Client: China Enterprise Co., Ltd.
Design scope: Proposal Design, Design Development
Design time: 2002
Completion time: 2006

总平面图 Site Plan

项目南侧的黄金城道是古北新区东西向重要的空间主轴，也是古北新区的景观中轴和公共空间，具有商业、娱乐功能。黄金城道的西南角设有步行入口广场，使区域文化休闲特征与社区高档的居住氛围得以相互交融，相互提升。

景观设计以绿化和水体相结合为主，采用局部对称和总体灵活的手法，很好地顺应了建筑本身所形成的空间特点，将建筑空间与景观设计融为一体。从会所周围流出的水流缓缓地汇成小溪，流向社区中心。水面、草坪、树木和铺地有机地组合在一起，带给住户优雅宁静、多样而统一的景观感受。

The Huangjincheng Avenue on the south side of the project is an important east-west spatial axis of the Gubei New Area and a landscape axis as well as a public space with commercial and entertainment functions. At the southwest corner of the Huangjincheng Avenue, there is a pedestrian entrance plaza so that the regional cultural and leisure features and the high-end residential atmosphere of the community can blend with and enhance each other.

The landscape design is mainly based on the combination of greening and water, and adopts a strategy of partial symmetry and overall flexibility. It conforms to the spatial characteristics of the building itself and integrates the architectural space and landscape design. The water flowing out from the clubhouse slowly merges into a stream and flows to the community centre, while the water surface, lawn, trees and paving are organically combined together to bring the residents an elegant, tranquil, diverse and unified landscape experience.

总用地面积：	4 hm²
总建筑面积：	110 000 m²
容积率：	2.8

Site area:	4 hm²
Gross floor area:	110,000 m²
F.A.R.:	2.8

总平面图 Site Plan

达安花园位于上海市静安区长寿路与武宁路交叉口。设计依据基地的方位条件，将住宅区分为三个组团，各个组团围绕各自的组团绿地（5 000~8 000 m²），三个组团又环绕一个约9 000 m²的中心花园，取得户户朝南，家家面绿的设计效果。空间归属感强，绿化生态环境优越，极佳地填补了现代城市生活所缺少的居住氛围。围合式空间的创造使住户在大型住区中更易于识别属于自己的生活空间，在视觉和心理上产生认同感和安全感。住户单元出入口均安排在组团花园周围，使花园更贴近日常生活。连接两个主出入口的车行道及带形花园是小区主要的东西轴线，具有重要的指向性和识别作用。

设计为充分体现长约200 m，宽约25 m的带形花园的生态功能，将原布置在带形花园中央的会所转移，使这里成为一条以绿树为主的生态林带，突出达安花园的生态主题。同时，绿色的轴线也为居民带来更多的享受。会所将替代花园中亭子等小品建筑的位置，并采用半地下的方式降低体量感。会所的玻璃体建筑成为花园的组成部分，既发扬了原有的绿色主题，又为会所的使用者提供了丰富的环境因素，使总体的设计价值得以提升。

Da'an Garden is located at the intersection of Changshou Road and Wuning Road in Jing'an District, Shanghai. The design is based on the site location. The residential area is divided into three groups. Each group surrounds its own green space (5,000~8,000 m²), the three groups encircle a central garden of about 9,000 m², achieving the design effect of households facing south and green. The strong sense of belonging to the space and the superior greening ecological environment perfectly complement the living atmosphere lacking in the modern urban life. The creation of enclosed space makes it easier for residents to identify their own living space in large-scale residential areas. And it produces a sense of identity and security visually and psychologically. The entrances of resident units are arranged around the group garden, making the garden closer to daily life. The roadway and the belt-shaped garden connecting the two main entrances are the main east-west axes, which have important directivity and identification.

The design is to fully reflect the ecological function of the belt-shaped garden with a length of about 200 m and a width of about 25 m. The clubhouse originally arranged in the centre of the belt-shaped garden is moved so that an ecological forest belt dominated by green trees highlights the ecological theme of Da'an Garden. Meanwhile, the green axis also inspires joy for the residents. The clubhouse replaces the pavilions and other small buildings in the garden, adopting a semi-underground method to reduce the sense of volume. The glass clubhouse becomes an integral part of the garden, which not only promotes the original green theme, but also provides abundant environmental factors for the users of the clubhouse, enhancing the overall design value.

总用地面积:	9 hm²
总建筑面积:	310 000 m²
容积率:	3.4
住宅总户数:	2 865

Site area:	9 hm²
Gross floor area:	310,000 m²
F.A.R.:	3.4
Total amount of units:	2,865

上海静安达安锦园
Da'an Jin Garden, Jing'an, Shanghai

地点：中国，上海
业主：上海达安锦迪置业有限公司
设计范围：方案设计、扩初设计
设计时间：2001
建成时间：2008

Location: Shanghai, China
Client: Shanghai Da'an Jindi Real Estate Co.
Design scope: Proposal Design, Design Development
Design time: 2001
Completion time: 2008

总平面图 Site Plan

设计在总体布局上，最大限度地利用城市绿地，同时又创造出优美的居住环境，使大部分住宅拥有优美的景观。项目南侧以及北侧均设有封闭式庭院，使住户能同时享受花园和城市景观。

环境设计沿用了上海的区域文化和先进的环境设计理念，创造出独特、美观、经济、实用的景观环境。

In terms of the overall layout, the design maximizes the use of urban green space and creates a beautiful residential environment at the same time, so that most of the houses are oriented to the beautiful landscape. The south and north sides of the project are designed to have an enclosed courtyard that gives each family a garden and an urban view in the meantime.

The environment design uses both Shanghai regional culture and modern environment design techniques to create a unique, beautiful, economical and practical landscape environment.

总用地面积:	4.3 hm²
总建筑面积:	160 000 m²
容积率:	3.7

Site area:	4.3 hm²
Gross floor area:	160,000 m²
F.A.R.:	3.7

上海松江达安圣芭芭花园河谷3号——90花园别墅
Da'an St. Babara Valley No.3—90 Villa, Songjiang, Shanghai

地点：中国，上海
业主：上海达安泰豪置业有限公司
设计范围：方案设计、扩初设计
设计时间：2007
建成时间：2010

Location: Shanghai, China
Client: Shanghai Da'an Taihao Real Estate Co., Ltd.
Design scope: Proposal Design, Design Development
Design time: 2007
Completion time: 2010

总平面图 Site Plan

本项目位于上海市松江区新桥镇。基地北邻明中路，西邻月台路，南邻春申塘，基地西部有南北向河道小茜浦流过。场地为狭长的长方形。

整体住房规划包括三层以下的联体别墅和少量独立住宅，入口处设有少量配套低层公建设施。

设计的主力户型是面积为90 m²以下的创新联体别墅。它不仅是住宅工程，同时也可以成为很人性、很优美、很舒适的结合室内外空间的优秀居住建筑典型。设计的主要原则是：所有屋顶以小坡屋顶为主，且不得相连；每户之间有后退，形成空间，并种植高大树木与竹子，在立面上形成独立的感觉；设计强调竖向线条、丰富的色彩和各不相同的相邻立面，体现独立性和个性；每层后退，形成平台。立面窗户被放大，符合当地的居住需求。设计包括建筑与环境，花坛、花钵、木艺、铁艺与建筑结合，相得益彰。建筑采用框架结构，抗震性能好，室内无承重墙。

"90花园别墅"是KFS公司的中国国家专利产品。

The site is located in Xinqiao Town in Songjiang District of Shanghai. The base is adjacent to Mingzhong Road in the north, Yuetai Road in the west, Chunshentang in the south, and Xiaoqianpu, a north-south river runs through the west of the base. The site is a long and rectangular shape.

The total housing program consists of 3-storey townhouses and a small number of detached houses. Low level public facilities are situated near the entrance.

The main unit of the design is an innovative townhouse with an area of less than 90 m². It is not only a housing project, but also a very human, beautiful, comfortable, and excellent example of residences that combine indoor and outdoor spaces. The main principles of the design are: all roofs are dominated by small slope roofs and must not be connected; there is a retreat between each household to form a space, and tall trees and bamboos are planted to form a sense of independence on the facade; the design emphasizes the vertical lines, the rich colours, and the different adjacent facades, reflecting independence and individuality; each floor retreats to form a platform. The facade windows are enlarged to meet the local residential needs. The building and the environment are designed together, and the flower beds, flower pots, wood art, and iron art are combined with the building to complement each other. The frame structure is adopted for a good seismic performance. There is no load-bearing wall indoors.

The "90 Villa" is a national patent product of KFS.

总用地面积：	4.3 hm²
总建筑面积：	30 000 m²
容积率：	0.7
住宅总户数：	268

Site area:	4.3 hm²
Gross floor area:	30,000 m²
F.A.R.:	0.7
Total amount of units:	268

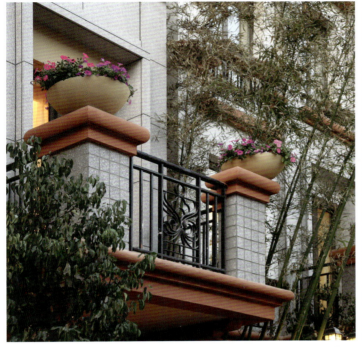

上海崇明达安御廷——90花园别墅
Da'an Royal Garden—90 Villa, Chongming, Shanghai

地点：中国，上海
业主：上海达安锦迪置业有限公司
设计范围：方案设计、扩初设计
设计时间：2007
建成时间：2010

Location: Shanghai, China
Client: Shanghai Da'an Jindi Real Estate Co., Ltd.
Design scope: Proposal Design, Design Development
Design time: 2007
Completion time: 2010

总平面图　Site Plan

项目位于上海崇明区城桥新城。规划为多层及低层住宅并配置少量社区服务用房。

根据规划，住宅被划分为四种不同的产品类型：多层、联排、独幢和双拼。"90花园别墅"是KFS公司的国家专利产品。

考虑到空间形态、交通组织和景观资源等因素，不同的住宅产品被由西向东依次成组团排布。西侧为多层住宅，东侧为独栋住宅，中间设置联排住宅，形成清晰的城市天际线。西侧多层住宅区的大面积集中绿化与东侧30 m绿化带相互呼应，使整个小区形成均衡有机的绿化体系。

The site is located in Chengqiao New Town, Chongming District of Shanghai. The plan incorporates multi-storey and low-rise units and a small number of community service units.

Based on the planning, the residence is divided into four different residential types—multi-storey buildings, townhouses, detached houses and semi-detached houses. The "90 Villa" townhouse is a national patent product of KFS.

Considering the factors of spatial forms, transportation systems and landscape resources, different residential types are placed in blocks form from west to east. There are multi-storey units on the west, and detached houses on the east. The townhouses are positioned in the middle. The skyline is clearly defined. The large-scale concentrated greenery of the multi-storey residential area on the west side and the 30 m green belt on the east side echo each other, forming a balanced organic green system for the entire community.

总用地面积：	8.2 hm²
地上建筑面积：	78 000 m²
容积率：	0.95
住宅总户数：	783

Site area:	8.2 hm²
Above ground G.F.A.:	78,000 m²
F.A.R.:	0.95
Total amount of units:	783

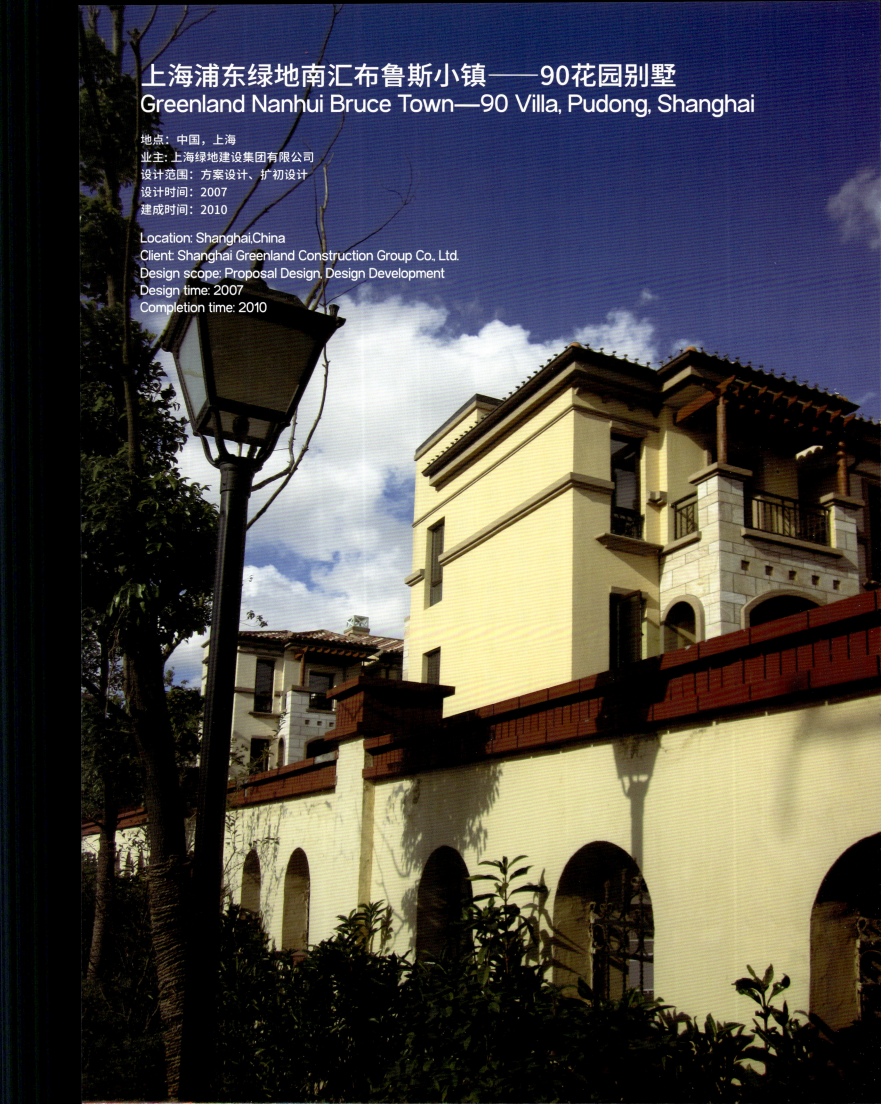

上海浦东绿地南汇布鲁斯小镇——90花园别墅
Greenland Nanhui Bruce Town—90 Villa, Pudong, Shanghai

地点：中国，上海
业主：上海绿地建设集团有限公司
设计范围：方案设计、扩初设计
设计时间：2007
建成时间：2010

Location: Shanghai,China
Client: Shanghai Greenland Construction Group Co., Ltd.
Design scope: Proposal Design, Design Development
Design time: 2007
Completion time: 2010

总平面图 Site Plan

本基地位于上海市南汇区惠南镇。基地北邻迎熏路，西邻浦东运河，南邻惠园路，东邻南团公路。老港河将地块分为南北两个部分。规划包括小高层及低层住宅、社区服务及商业用房。

设计将本项目定位为一座都市田园小镇。不同户型的住宅组合形成复合型的社区围氛。花园式的居住环境以及开放性和私密性的景观，形成安宁、放松的氛围。

联排产品为"90花园别墅"——KFS公司的国家专利产品。

总用地面积：	8.6 hm²
总建筑面积：	106 800 m²
容积率：	1.3
住宅总户数：	721

The site is located in Huinan Town in Nanhui District of Shanghai and is adjacent to Yingxun Road on the north, Pudong Canal on the west, Huiyuan Road on the south, and Nantuan Highway on the east. The Laogang River divides the land into the north and south parts. The master plan consists of mid-rise and low-rise residential units, service facilities and retail blocks.

The design defines the project as an urban garden town. The combination of units with different sizes is conducive to a mixed social atmosphere. The garden-like environment with public and private landscapes results in a peaceful and relaxed atmosphere.

The "90 Villa" townhouse is a national patent product of KFS.

Site area:	8.6 hm²
Gross floor area:	106,800 m²
F.A.R.:	1.3
Total amount of units:	721

上海黄浦明日星城
Tomorrow Star City, Huangpu, Shanghai

地点：中国，上海
业主：上海东方金马房地产发展有限公司
设计范围：方案设计、扩初设计（部分一期除外）
设计时间：2001
建成时间：2005

Location: Shanghai, China
Client: Shanghai Orient Golden Horse Co., Ltd.
Design scope: Proposal Design, Design Development (Exclude Part of Phase 1)
Design time: 2001
Completion time: 2005

总平面图 Site Plan

明日星城位于上海市黄浦区老城区中心，整个基地被城市道路分隔成六个街区。城市型的住宅区与商业、影视娱乐、教育文化等多种功能相结合。设计力求创造出高尚的生活环境。总体布局在极为紧凑的用地内尽可能多地布置了休闲绿地，以营造闹中取静的居住氛围。沿街住宅底部设商店并利用商业裙房将小区内部新绿地围合成封闭式的内院，形成一个与喧闹的街道隔离开来的安静的居住空间。建筑主要以18-35层的高层住宅为主，不同层数相互组合，使建筑体量呈现出变化丰富的外观效果。

Tomorrow Star City is located in the centre of the old area of Huangpu District, Shanghai. The entire base is divided by urban roads into six blocks. The urban-type residential area integrates commerce, film and television entertainment, education and culture along with other functions. The design strives to create a noble living environment. In the overall layout, there are as many leisure green spaces as possible in the extremely compact land to create a quiet living atmosphere along the noisy road. Shops are set up at the bottom of the residential buildings along the street and commercial podiums are used to enclose the new green space in the community into a closed inner courtyard, forming a quiet residential space isolated from the noisy street. The buildings are mainly high-rise residential buildings with 18 to 35 floors, and different layers are combined to make the building volume present a varied appearance effect.

总用地面积：	11.1 hm²
总建筑面积：	450 000 m²
容积率：	4.0

Site area:	11.1 hm²
Gross floor area:	450,000 m²
F.A.R.:	4.0

Location: Shanghai, China
Client: Shanghai Dahua Group Co., Ltd.
Design scope: Proposal Design, Design Development
Design time: 2002
Completion time: 2007

总平面图 Site Plan

愉景华庭位于上海市普陀区。设计根据40 m宽的中央绿化带的规划特色，塑造出宜人的居住环境。

设计沿中央绿化带布置大户型，沿东西干道布置小高层建筑，从而与内部高层建筑形成对比，突出中央绿化带。

The Yujing Garden in Putuo District, Shanghai has a 40 m wide central green belt as the planning feature to create a comfortable living environment.

The design arranges large units along the central green belt and arranges mid-rise buildings along the east-west trunk road to contrast with the internal high-rise buildings, thereby highlighting the central green belt.

总用地面积:	9.1 hm²
总建筑面积:	227 000 m²
容积率:	2.1
住宅总户数:	1 630

Site area:	9.1 hm²
Gross floor area:	227,000 m²
F.A.R.:	2.1
Total amount of units:	1,630

上海长宁春天花园
Spring Garden, Changning, Shanghai

地点：中国，上海
业主：上海东方金马房地产发展有限公司
设计范围：方案设计、扩初设计
设计时间：2001
建成时间：2005

Location: Shanghai, China
Client: Shanghai Orient Golden Horse Co., Ltd.
Design scope: Proposal Design, Design Development
Design time: 2001
Completion time: 2005

总平面图 Site Plan

春天花园位于上海市长宁区长宁路，靠近虹桥开发区。设计以错位经营的思维模式为指导，创造了与周边地块性质相同而产品定位不同的新型住宅小区。整个住宅区域分为5层的叠加别墅与18~30层的小高层及高层。不同类型住宅带来的建筑体型变化，丰富了长宁路主要干道沿线的城市天际线，获得了城市规划部门和消费者的肯定。

叠加别墅位于地块南部，高层住宅位于地块北部，带来最大化的日照和对中央公园良好的视线。主入口大道和中央公园结合了西方当代建筑特色和中国传统建筑特色。树木、草地、小溪和喷泉提供了丰富多样的游园体验。靠近公路一侧的首层被抬高，将花园纳入城市的景观视线中，使其成为长宁区独特的城市设计体验。

Spring Garden is located on Changning Road, Changning District, Shanghai, nearing Hongqiao Development Zone. The design is guided by the thinking mode of dislocation management, creating a new type of residential community with the same nature as the surrounding land but with differentiated product positioning. The entire residential area is divided into 5-storey superimposed villas and 18 to 30-storey mid-rise and high-rise buildings. The changes in building shapes brought about by different types of residences have enriched the city's skyline along the main artillery of Changning Road, and won the affirmation of the urban planning department and the consumers.

The superimposed villas are located in the southern part of the site, and the high-rise residential buildings are located in the northern part, bringing maximum sunshine and a good view of the central park. The main entrance avenue and the central park combine contemporary western and traditional Chinese architectural features. Trees, lawns, creeks and fountains provide a rich and varied garden experience. The first floor on the side of the road is raised to bring the garden into the sight of the city, making it a unique urban design experience in Changning District.

总用地面积:	6.5 hm²
总建筑面积:	160 000 m²
容积率:	2.5
住宅总户数:	1 177

Site area:	6.5 hm²
Gross floor area:	160,000 m²
F.A.R.:	2.5
Total amount of units:	1,177

上海虹口明佳花园
Mingjia Garden, Hongkou, Shanghai

地点：中国，上海
业主：上海明佳房产开发有限公司
设计范围：方案设计、扩初设计
设计时间：2001
建成时间：2005

Location: Shanghai, China
Client: Shanghai Mingjia Real Estate Co., Ltd.
Design scope: Proposal Design, Design Development
Design time: 2001
Completion time: 2005

总平面图 Site Plan

明佳花园位于上海市虹口区的中心，靠近四川北路商业街及新建城市公园。

虹口区文化气息浓厚。小区在总体布局上将更多的住宅单元面向四川北路的绿地，使大部分住宅既有最佳朝向，又有最美景观。结合重点立面的处理，立面成为建筑群的视觉焦点。

小区环境最大限度地借用四川北路的绿地景观，把住宅区景观设计融入大面积的绿地中，使两者互为补充与延伸。在建筑的立面上，设计打破高层板式建筑固有的巨大体量。立面处理分散了体量感，减轻了建筑顶部的重量感，使建筑形象轻盈、体量变化丰富。设计在开敞的绿地旁制出一种轻松、优美的居住建筑形象。同时，建筑细部尽量运用玻璃、轻钢等装饰以减轻大型建筑体量带来的视觉上的压抑感。

Mingjia Garden is located at the centre of Hongkou District of Shanghai, adjacent to Sichuan North Road commercial street and the newly built urban park.

Hongkou District has a strong cultural atmosphere. In terms of the overall layout of the community, more residential units are designed to face the green space of Sichuan North Road. Most residences have the best orientation and the most beautiful views. Combined with the key facade treatment, the facade becomes the visual focus of the building complex.

The community environment borrows the green landscape of Sichuan North Road to the greatest extent, and integrates the residential landscape into a larger green space, so that the two complements can supplement each other. On the facade of the building, the design breaks the huge volume of high-rise slab buildings. The facade treatment disperses the volume, reduces the weight of the building's roof, and makes the building lighter and richer in volume changes. The design creates a relaxing and beautiful residential image near the open green land. Simultaneously, the architectural details are decorated with glass and light steel as much as possible to weaken the visual depression brought by the large building volume.

总用地面积:	2.5 hm²
总建筑面积:	100 000 m²
容积率:	4.0
总户数:	667

Site area:	2.5 hm²
Gross floor area:	100,000 m²
F.A.R.:	4.0
Total amount of units:	667

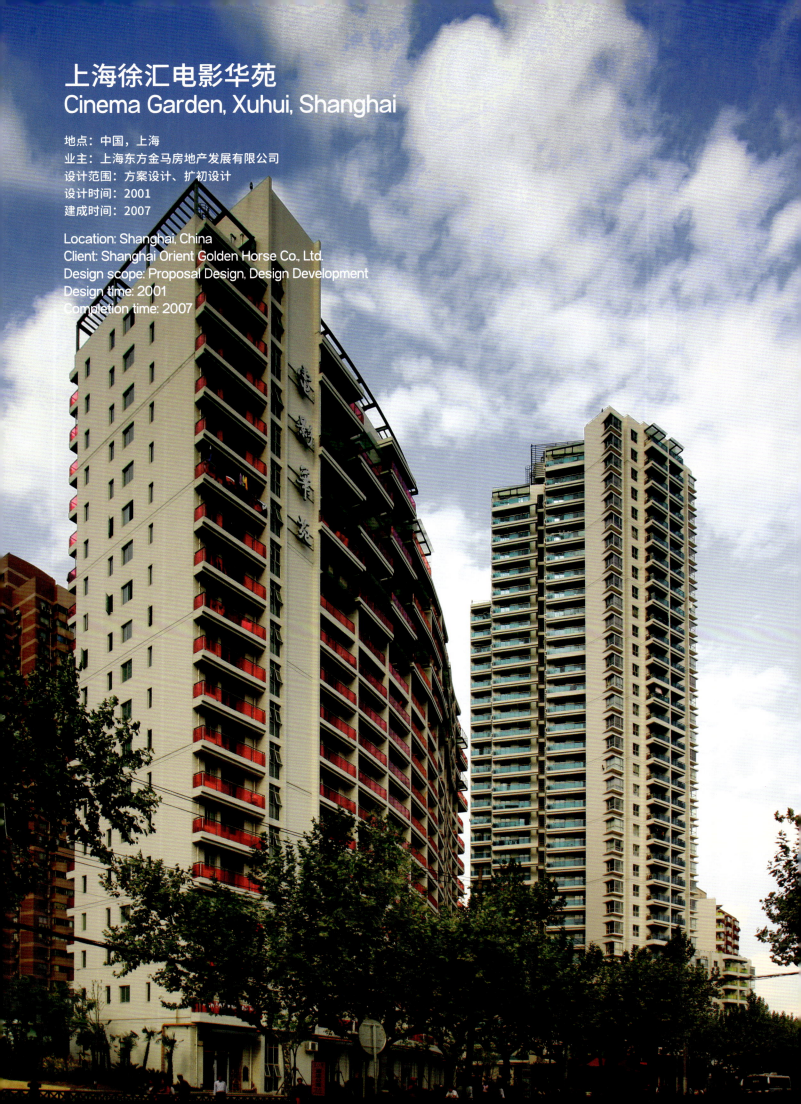

上海徐汇电影华苑
Cinema Garden, Xuhui, Shanghai

地点：中国，上海
业主：上海东方金马房地产发展有限公司
设计范围：方案设计、扩初设计
设计时间：2001
建成时间：2007

Location: Shanghai, China
Client: Shanghai Orient Golden Horse Co., Ltd.
Design scope: Proposal Design, Design Development
Design time: 2001
Completion time: 2007

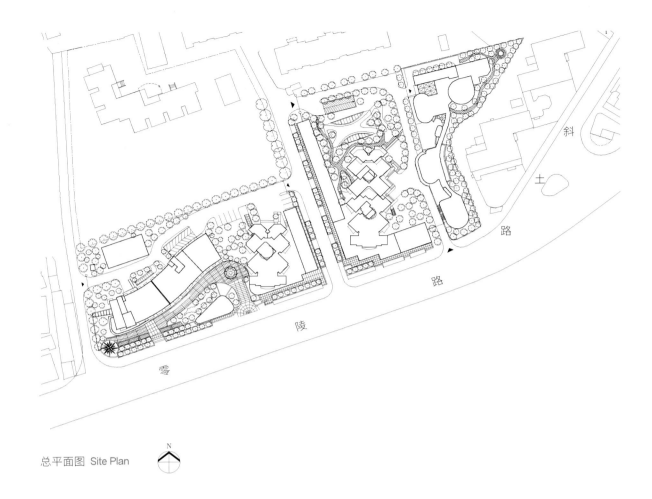

总平面图 Site Plan

本项目基地原为上海电影制片厂。设计尽可能地与周边环境相协调，形成唯一的建筑布局形式。房型以60~80 m²的小户型和出租性公寓为主，并设有一个外向型的小型商业步行街。会所设在地下室，包括一个室内恒温游泳池。这些配套设施将极大地满足住户的日常需求。

高低错落的建筑布局既受基地日照条件的限制，也给城市带来丰富的天际效果。地下停车库顶板低于路面0.6 m。设计堆土造景，组织跌水涌泉，结合绿篱、草地、雕塑等共同塑造出优美的户外生活空间。每套住宅均力求创造明亮整洁、舒适宜人的居住空间。

建筑的立面采用大量玻璃栏板阳台及大面积低窗台凸窗。沿街商业裙房设有一系列精致的挑篷，充分体现了简洁、优雅的建筑风格。

The project is in the former site of the Shanghai Film Studio. The design aims to coordinate the surrounding environment with every possibility to form a unique architectural layout form. The room types are mainly small and rental apartments of 60 to 80 m². There is also an outward-oriented small commercial street. The clubhouse is located in the basement and includes an indoor heated swimming pool. These facilities will greatly meet the daily needs of the residents.

The layout of the buildings is defined by sunlight conditions that determine different heights, which in turn creates a beautiful skyline to the city. The underground garage roof is 0.6 m lower than the road surface. By mounding the earth, the design creates a beautiful outdoor living space. It organizes water and springs and combines hedges, grass, sculptures, and more. Each residence strives to create a bright, clean, comfortable and pleasant living space.

The facades of the buildings widely adopt glass fence balconies and large-area low bay windows. The commercial podiums along the street have a series of exquisite canopies, which fully reflects the simple and elegant architectural style.

总用地面积:	1.9 hm²
总建筑面积:	66 000 m²
容积率:	3.5

Site area:	1.9 hm²
Gross floor area:	66,000 m²
F.A.R.:	3.5

上海徐汇漕河景苑
Caohejing Garden, Xuhui, Shanghai

地点：中国，上海
业主：上海汇成房产经营公司
设计范围：方案设计、扩初设计
设计时间：2002
建成时间：2007

Location: Shanghai, China
Client: Shanghai Huicheng Real Estate Co., Ltd.
Design scope: Proposal Design, Design Development
Design time: 2002
Completion time: 2007

总平面图 Site Plan

漕河景苑位于上海徐汇区。小区以住宅为主，辅以商业、会所等配套设施。基地南侧、东侧为宽阔水域，具有一定的景观优势。基地所处区域环境较为幽静，整体环境优良，周边交通发达，出行方便。

设计将自然水面与人工水面巧妙结合。设计还将建筑的景观与水景结合，使得小区成为城市的一道亮丽风景线，达到城市规划需求与住宅开发效益的双向共赢。基地东部是青年公寓，景观极其开阔，立面造型挺拔、隽秀，成为半岛形基地上的地标式建筑。

设计沿河岸布置了两排精致的低层住宅。西侧的高层住宅与会所映衬出美丽的沿江风景。河的南侧则由绿岛、沙滩与水生植物组成了生态景观。小区的中心是以自然手法营造的中心绿地，使该组团住宅可享受南北双向景观。小区的北部是有便利商业设施的南向大纵深观景房。

The Caohejing Garden is located in Xuhui District, Shanghai. The community is mainly for residential development, supplemented by facilities such as commerce and clubs. The south and east sides of the site are wide waters, which have certain landscape advantages. The environment of the area where the site is located is relatively quiet, with excellent environment, developed surrounding traffic, and convenient transportation.

The design integrates the natural waterscape with the man-made waterscape. The design also combines the architectural landscape with the waterscape, making the community a beautiful landscape for the city, achieving a win-win situation between the needs of urban planning and the benefits of residential development. The eastern part of the site lies a youth apartment with an extremely open landscape. With the tall and straight facade, it becomes a landmark building on the peninsula-shaped site.

The design arranges two rows of exquisite low-rise residential buildings along the river bank. The high-rise residential buildings and the clubhouse on the west side set off the beautiful scenery along the river. On the south side of the river, green islands, beaches and aquatic plants form an ecological landscape. The centre of the residential area is a central green space created by natural methods, so that the group of residential buildings can enjoy the landscape of north and south. To the north of the community is a south-facing large and deep viewing room with convenient commercial facilities.

总用地面积：	4.5 hm²
总建筑面积：	113 000 m²
容积率：	2.5

Site area:	4.5 hm²
Gross floor area:	113,000 m²
F.A.R.:	2.5

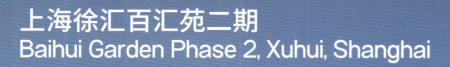

上海徐汇百汇苑二期
Baihui Garden Phase 2, Xuhui, Shanghai

地点：中国，上海
业主：上海百汇房地产开发有限公司
设计范围：方案设计、扩初设计
设计时间：2007
建成时间：2013

Location: Shanghai, China
Client: Shanghai Baihui Real Estate Development Co., Ltd.
Design scope: Proposal Design, Design Development
Design time: 2007
Completion time: 2013

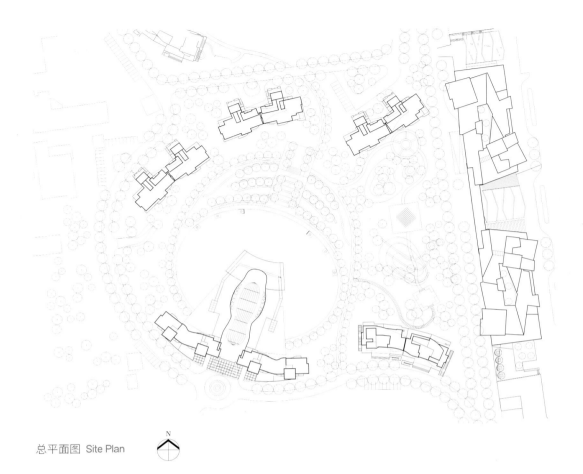

总平面图 Site Plan

此项目为百汇苑二期的住宅部分，设计在进行户型优化的同时进行立面改造设计。

本项目地处上海市徐汇区南部，东邻黄浦江，北靠龙华港，并与2010年上海世博会场址隔江相对，得天独厚的景观优势是本项目的最大潜力所在。因此，景观资源的最大化利用成为此次方案调整的指导方针。

在这个项目中，设计师们提出舒适观景的新概念。例如，坐在客厅的沙发上或者躺在主卧室的床上即能尽览江面壮丽的美景。要达到这一效果，设计就必须在房型设计上充分考虑户型和景观资源的相对关系，公共空间和私密空间的位置以及开窗的方式和阳台的位置、观景面的幅度等很多细节。

| 总用地面积： | 21.7 hm² |
| 总建筑面积： | 165 000 m² |

This project is to design the residential parts in Phase 2 of Baihui Garden, including facade renovation as well as optimizing accommodation units.

The project is located in the south of Xuhui District, Shanghai, adjacent to Huangpu River in the east and Longhua Port in the north, and facing the 2010 World Expo site across the river. The unique landscape advantage is the greatest potential of this project. Therefore, the maximum utilization of landscape resources has become the guiding principle of this program adjustment.

In this project, the designers put forward a new concept of comfortable viewing. For example, sitting on sofa in the living room or lying in bed in the master bedroom, residents can enjoy the magnificent view of the rive. To achieve it, the design is necessary to fully consider the relationship between house types and landscape resources, the location of public and private spaces, window openings, location of balconies, and the range of viewing areas, etc., in room layouts.

| Site area: | 21.7 hm² |
| Gross floor area: | 165,000 m² |

总平面图 Site Plan

上海滩花园洋房坐落于上海浦东陆家嘴金融贸易区内，交通便捷，生活、娱乐、商业设施完备，是低容积率的高档住宅小区。

为满足居住者亲近自然的愿望，建筑采用了层层退台的形式，创造出自然的体验。建筑体量有效地缩短了人与自然的距离。住宅完全掩映在绿树花香之中，体现了都市园林的设计意境。

Shanghaitan Garden House is located in the Lujiazui Financial and Trade Zone in Pudong, Shanghai. It has convenient transportation, complete living, entertainment, and commercial facilities. It is a high-end residential community with low F.A.R.

In order to satisfy the residents' desires to be close to nature, the building adopts the form of step-back terraces, which creates a natural experience. The building volume effectively shortens the distance between man and nature. The houses are completely hidden in the green trees and the scent of flowers, reflecting the artistic conception of urban garden design.

总用地面积: 14.4 hm²
总建筑面积: 136 000 m²
容积率: 0.94
住宅总户数: 635

Site area: 14.4 hm²
Gross floor area: 136,000 m²
F.A.R.:   0.94
Total amount of units: 635

上海浦东东晶国际公寓
Dongjing International Residence, Pudong, Shanghai

地点：中国，上海
业主：上海东道置业有限公司
设计范围：方案设计、扩初设计
设计时间：2002
建成时间：2007

Location: Shanghai, China
Client: Shanghai Dongdao Real Estate Co., Ltd.
Design scope: Proposal Design, Design Development
Design time: 2002
Completion time: 2007

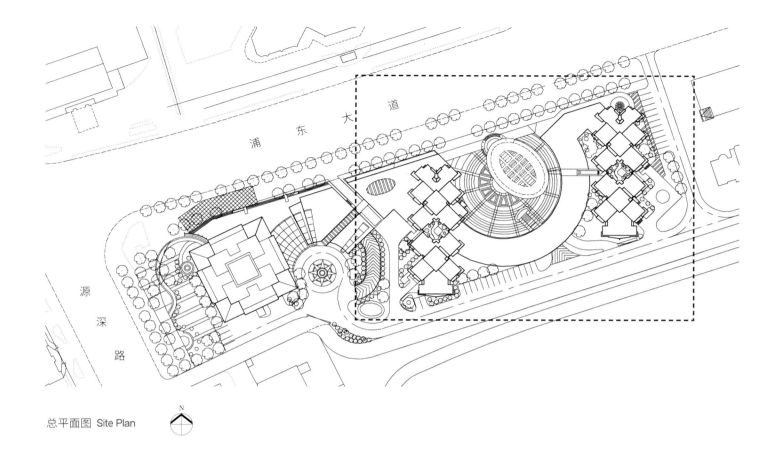

总平面图 Site Plan

该基地位于上海浦东大道，包含居住和商业等功能。居住和商业功能被细化以实现"双赢"。设计采用新的建筑形式，创造出一种新的国际化、多功能综合体。曲面的商业建筑集中布置在两栋公寓楼之间，沿街布置以中小型复式商铺为主的连续商业，有效围合形成具有园林景观特色的外向型城市商业广场。

本方案意在从建筑文化和环境人文角度着手，用崭新的时代建筑语言，建设一个环境优美、设施完备、风格独特的国际化居住园区。多种功能既合又分的布局实现多赢格局。简约并极富韵律感的总体布局以及戏剧性的构图，使得空间层次更加丰富。现代的建筑外观及商业建筑明亮、通透的立面造型，在浦东大道上让出更多的空间，留下更多的阳光，使其对城市主干道的压力大为减少并提升了内部空间品质，彰显小区的国际化概念。

The site is situated on Pudong Avenue, Shanghai, and includes residential and commercial functions. Residential and commercial functions have been refined to achieve a "win-win" situation. The design adopts a new architectural form, creating a new international and multifunctional complex. Curved commercial buildings are centrally arranged between two apartment buildings, and continuous commerce mainly composed of small and medium-sized duplex shops are arranged along the street, effectively enclosing and merging to form an outward-oriented urban commercial plaza with garden landscape characteristics.

This plan intends to build an international residential park with a beautiful environment, complete facilities and a unique style from the perspectives of architectural culture and environmental humanities, using a brand-new architectural language of the times. The combined and separated layout of multiple functions realizes a win-win situation. The simple and rhythmic overall layout and dramatic composition make the spatial layers richer. The appearance of modern buildings and the bright and transparent facades of commercial buildings make more space on Pudong Avenue and leave more sunlight, which greatly reduces the pressure on the city's main roads and improves the quality of the interior space. It also highlights the concept of internationalization of the community.

总用地面积：	1.7 hm²
总建筑面积：	66 000 m²
容积率：	3.8
公寓总建筑面积：	33 000 m²
住宅总户数：	452

Site area:	1.7 hm²
Gross floor area:	66,000 m²
F.A.R.:	3.8
G.F.A. of apartment:	33,000 m²
Total amount of units:	452

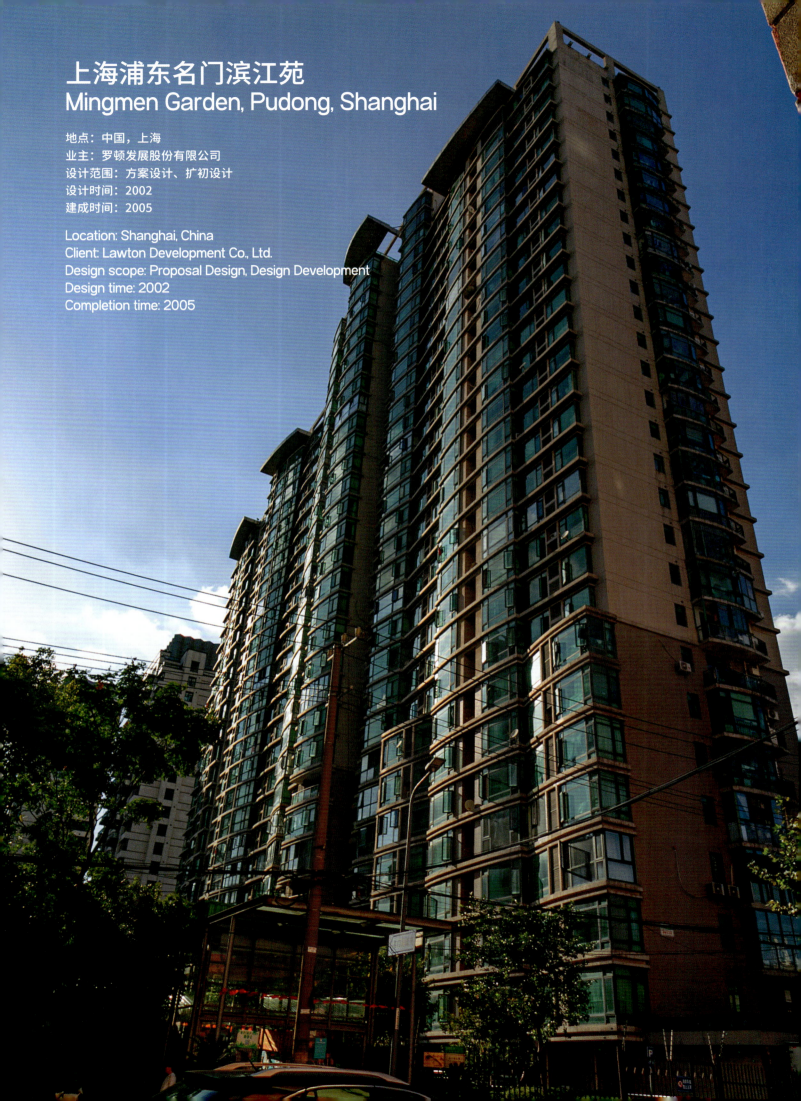

上海浦东名门滨江苑
Mingmen Garden, Pudong, Shanghai

地点：中国，上海
业主：罗顿发展股份有限公司
设计范围：方案设计、扩初设计
设计时间：2002
建成时间：2005

Location: Shanghai, China
Client: Lawton Development Co., Ltd.
Design scope: Proposal Design, Design Development
Design time: 2002
Completion time: 2005

本项目位于浦东大道北侧，铜山路西侧，北眺黄浦江。内院空间由一栋高层、一栋小高层与一栋多层围合而成。设计的目标是在保持各种功能，满足各种规范的基础上对布局进行修改，使其完整、充实、优雅宜人。

从小区入口广场开始，一条玻璃连廊连接了1号楼的两个门厅；另一条玻璃连廊勾勒出一条优美的曲线，通向2号楼，环抱一个约600 m²的水池。远处的景观瀑布是池水的源头，使住户在远眺黄浦江之时，心中更有亲切感。

The project is located on the north of Pudong Avenue and west of Tongshan Road, with clear views of Huangpu River towards the north. The inner yard is embraced by a high-rise building, a medium height building and a multi-storey building. The purpose of the design is to modify the layout to make the form complete, fulfilling and elegant, all the while maintaining the functions and requirements.

Starting from the entrance square of the community, a glass corridor connects the two entrance halls of Building 1, while the other outlines a graceful curve leading to Building 2 and surrounds a pool of about 600 m². The scenic waterfall in the distance is the source of the pool water, which makes residents feel more genial in their hearts when they overlook Huangpu River.

总用地面积：	0.46 hm²
总建筑面积：	10 111 m²
容积率：	2.2
住宅总户数：	96
绿化率：	40.20%

Site area:	0.46 hm²
Gross floor area:	10,111 m²
F.A.R.:	2.2
Total amount of units:	96
Rate of greening:	40.20%

上海浦东东南华庭
Dongnan Garden, Pudong, Shanghai

地点：中国，上海
业主：罗顿发展股份有限公司
设计范围：方案设计、扩初设计
设计时间：2004
建成时间：2006

Location: Shanghai, China
Client: Lawton Development Co., Ltd.
Design scope: Proposal Design, Design Development
Design time: 2004
Completion time: 2006

本项目位于浦东南路与临沂北路的东侧，由两座高层住宅组成，建筑造型典雅、明快。

The project is located in the east of Pudong South Road and Linyi North Road. It is composed of two high-rise residential buildings with elegant and lively architectural style.

总用地面积：	3.9 hm²
总建筑面积：	90 651 m²
容积率：	2.29
住宅总户数：	856

Site area:	3.9 hm²
Gross floor area:	90,651 m²
F.A.R.:	2.29
Total amount of units:	856

珠海横琴华发广场
Huafa Square, Hengqin, Zhuhai

地点：中国，珠海
业主：珠海华发房地产开发有限公司
设计范围：方案设计
设计时间：2017
建成时间：2018

Location: Zhuhai, China
Client: Zhuhai Huafa Real Estate Development Co., Ltd.
Design scope: Proposal Design
Design time: 2017
Completion time: 2018

本项目位于珠海横琴的东侧，与澳门隔水相望，是珠海的地标建筑，对于城市景观有着重要的影响。整体的设计以荷花的形态为概念，充分考虑第五立面设计，使设计成为城市景观，并提升城市形象。

This project is located on the east side of Hengqin, Zhuhai, facing Macao across the water. It is a landmark building in Zhuhai and has an important impact on the urban landscape. The overall design takes the shape of a lotus as the concept, and fully considers the design of the fifth facade, which becomes an urban landscape and enhances the image of the city.

总用地面积：	3.3 hm^2
总建筑面积：	99 000 m^2
容积率：	3.0

Site area:	3.3 hm^2
Gross floor area:	99,000 m^2
F.A.R.:	3.0

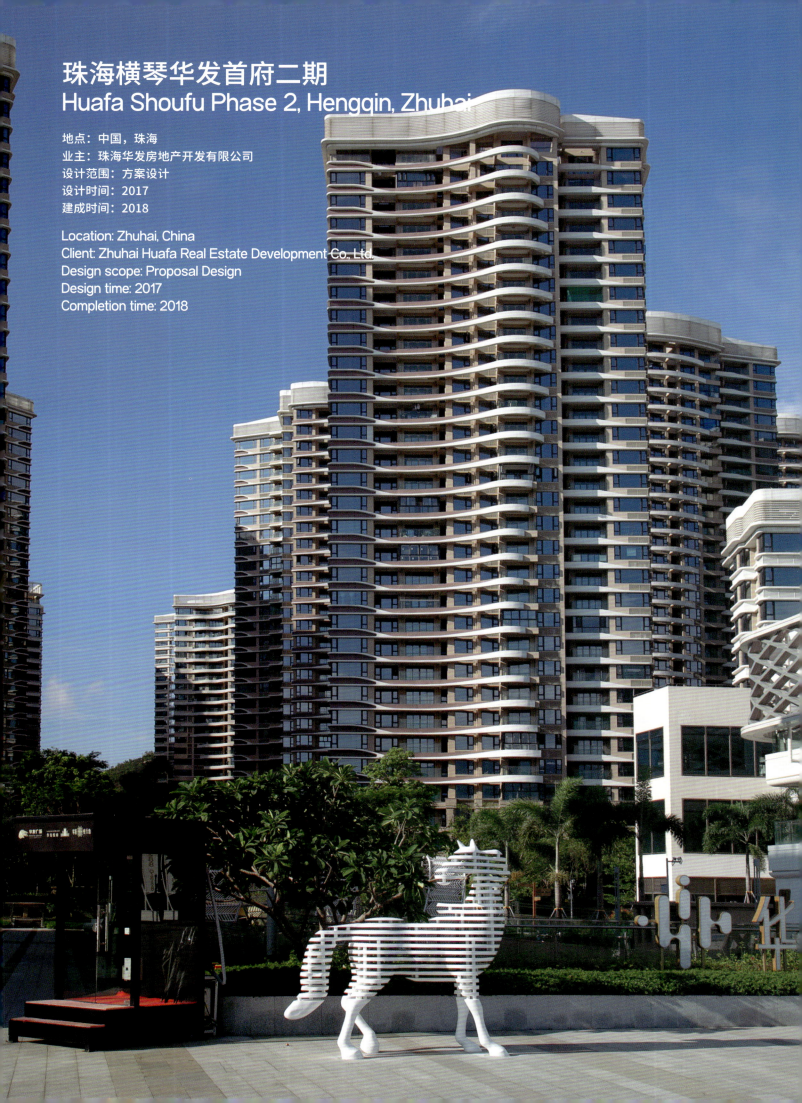

珠海横琴华发首府二期
Huafa Shoufu Phase 2, Hengqin, Zhuhai

地点：中国，珠海
业主：珠海华发房地产开发有限公司
设计范围：方案设计
设计时间：2017
建成时间：2018

Location: Zhuhai, China
Client: Zhuhai Huafa Real Estate Development Co. Ltd.
Design scope: Proposal Design
Design time: 2017
Completion time: 2018

整个小区采用人车分流的模式，车行交通完全在小区外部及地下。小区内部通过景观连廊通向每家每户，形成了私密的庭院空间。

立面设计增加曲线以及适量深色线条，强调白色飘带，从而让立面显得更加轻盈、美观。

入口大堂的设计强调了仪式感，并使用了石材大门。设计凸显了入口大堂空间，增加了入口空间的过渡层次，从而提升了居住空间的品质。

The community is designed to separate vehicle traffic from pedestrian traffic. Vehicle traffic is located underground and outside the community. Inside the community, each family is connected with one another through landscape corridors, forming private courtyards.

The facade design increases curves and multiple dark coloured lines while emphasizing white ribbons to make the facade light and beautiful.

The design of the entrance lobby emphasizes a sense of ritual by using the stone gate. The design draws attention to the space of the entrance lobby, which adds a transition level to the entrance space and improves the quality of the living space.

总用地面积：	5.3 hm²
总建筑面积：	133 192 m²

Site area	5.3 hm²
Gross floor area:	133,192 m²

珠海横琴华发首府三期
Huafa Shoufu Phase 3, Hengqin, Zhuhai

地点：中国·珠海
业主：珠海华发房地产开发有限公司
设计范围：方案设计、扩初设计
设计时间：2017
建成时间：2019

Location: Zhuhai, China
Client: Zhuhai Huafa Real Estate Development Co., Ltd.
Design scope: Proposal Design, Design Development
Design time: 2017
Completion time: 2019

"人的活动，是商业的真正立面。"

设计考虑水面及玻璃反射的光线影响，植入灰空间，形成空间过渡。

华发首府三期项目位于横琴天沐河中心区位。设计进行了整体构思，与周边项目相协调，打造出天沐河沿河景观带。

项目周边有游艇俱乐部、天心湾（克拉码头）等重要项目。

"The activities of people are the real facades of business."

The design considers the effects of light reflecting on water and glass, with the implementation of grey space, to form a spatial transition.

Huafa Shoufu Phase 3 project is located in the centre of Hengqin Tianmu River. The design takes overall layout into consideration and incorporates a landscape belt along the river, coordinating with the surrounding context.

The project is surrounded by important projects such as the yacht clubs and Tianxin Bay (Kela Marina).

总用地面积：	1.9 hm²
总建筑面积：	42 000 m²
容积率：	2.2

Site area	1.9 hm²
Gross floor area:	42,000 m²
F.A.R.:	2.2

武汉东西湖恋湖家园
Lakeville Garden, Dongxihu, Wuhan

地点：中国，武汉
业主：武汉深江物业发展有限公司
设计范围：方案设计、扩初设计
设计时间：2005
建成时间：2019

Location: Wuhan, China
Client: Wuhan Shenjiang Real Estate Co., Ltd.
Design scope: Proposal Design, Design Development
Design time: 2005
Completion time: 2019

每一块土地都有其本身的个性。规划设计的首要任务就是找到符合地块的规划模式，从而使土地的价值潜力得到挖掘。

基地位于武汉市东西湖区金银湖生态旅游区内，拥有800 m长的湖岸线。基地狭长的形状，使得全湖景住宅成为可能。

为了实现全湖景住宅这一构想，高层住宅呈"V"字形排列，开口朝向湖面。联排别墅也被布置在基地的邻湖处，也呈"V"字形排列。这样，组团内93%的住户都能看到湖面。

Every piece of land has its own characteristics. The priority of planning and design is to seek for the development pattern that fits the land, and attributes to reach its potential value.

The site is in the eco-tourist zone of Jinyin Lake in Dongxihu District, Wuhan. The site has a 800 m long shoreline. The linear condition of the site makes it possible for the residences to have uninterrupted lake views.

In order to realize the concept for the residences to have uninterrupted lake views, the high-rise buildings are arranged in a "V" shape with openings towards the lake. Townhouses are also arranged in a "V" shape close to the vicinity of the waterfront. Therefore, 93% of the residences in the group is designed with lake views.

总用地面积：	11.1 hm²
总建筑面积：	214 600 m²
容积率：	1.93
住宅总户数：	1 498
绿化率：	22.5%

Site area:	11.1 hm²
Gross floor area:	214,600 m²
F.A.R.:	1.93
Total amount of units:	1,498
Rate of greening:	22.5%

成都武侯上海花园
Shanghai Garden, Wuhou, Chengdu

地点：中国，成都
业主：成都新东方置业有限公司
设计范围：方案设计、扩初设计
设计时间：2004
建成时间：2010

Location: Chengdu, China
Client: Chengdu New Orient Real Estate Co., Ltd.
Design scope: Proposal Design, Design Development
Design time: 2004
Completion time: 2010

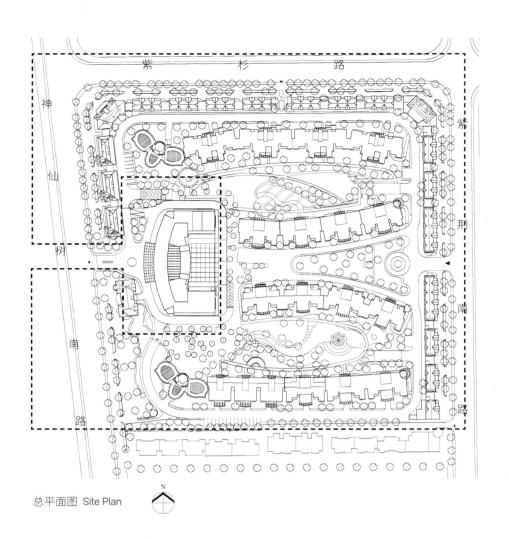

总平面图 Site Plan

上海花园是建于成都市高新区的高档住宅小区。基地东靠紫荆南路，西靠神仙树南路，南邻机场路西延线，北为紫杉路。基地内的五星级酒店大大提高了本小区住宅的价值。

Shanghai Garden is a high-end residential community built in Chengdu High-tech Industrial Development Zone. The site is adjacent to Zijing South Road in the east, Shenxianshu South Road in the west, west extension line of Airport Road in the south, and Zishan Road in the north. The five-star hotel in the site greatly increases the value of the residential quarters in the community.

总用地面积：	5.8 hm²
总建筑面积：	145 000 m²
容积率：	2.5

Site area:	5.8 hm²
Gross floor area:	145,000 m²
F.A.R.:	2.5

总平面图 Site Plan

两个车行出入口位于南侧与西侧。同时，南侧还有一个人行出入口，通过水池和铺地引领着人们进入院落之中。花园和底层架空空间使室内与室外空间相融合。自然景观轴线与规划景观轴线创造出宁静的气氛。

Two vehicle entrances are on the south and west. Meanwhile, a pedestrian entrance is also on the south, leading people into different courtyards with pools and pavements. The gardens and the raised ground floor space fusion the boundary of interior and exterior space. The natural landscape axis and the planned landscape axis generate a calm atmosphere.

总用地面积：	12.1 ha
总建筑面积：	290 000 m²
容积率：	2.4
住宅总户数：	1 718

Site area:	12.1 ha
Gross floor area:	290,000 m²
F.A.R.:	2.4
Total amount of units:	1,718

成都锦江上海东韵
Shanghai Dongyun, Jinjiang, Chengdu

地点：中国，成都
业主：成都新东方置业有限公司
设计范围：方案设计、扩初设计
设计时间：2004
建成时间：2012

Location: Chengdu, China
Client: Chengdu New Orient Real Estate Co., Ltd.
Design scope: Proposal Design, Design Development
Design time: 2004
Completion time: 2012

　　成都上海东韵是一个大型城市综合体，包括超高层的办公楼和五星级酒店，大型的会议、商业设施和高端住宅产品。

　　在项目的设计过程中，设计不仅考虑建筑形体和平面布局，还将设计纳入建造系统中进行综合考虑。设计更加注重成熟技术，注重建筑技术的广泛运用，特别是可能会引起后期隐患的构造，以及防水和材料的耐久性等问题。

Shanghai Dongyun is a large scale urban complex, including super high-rise offices and a five-star hotel, large conference and commercial facilities and high-end residential units.

During the design process of the project, the design considers the architectural form and layout as well as the construction system comprehensively. The design focuses more on proven technologies, the general use of building technologies, especially future potential structural defects, and the waterproofing and durability of materials.

总用地面积：	17.5 ha
总建筑面积：	400 000 m²
容积率：	2.5
住宅总户数：	3 000
绿化率：	40.5%

Site area:	17.5 ha
Gross floor area:	400,000 m²
F.A.R.:	2.5
Total amount of units:	3,000
Rate of greening:	40.5%

青岛城阳泰晤士小镇
Thames Town, Chengyang, Qingdao

地点：中国，青岛
业主：鲁商置业股份有限公司
设计范围：方案设计、扩初设计
设计时间：2013
建成时间：2017

Location: Qingdao, China
Client: Lushang Property Co., Ltd.
Design scope: Proposal Design, Design Development
Design time: 2013
Completion time: 2017

项目位于青岛市惜福镇，紧邻崂山景区，即墨区、城阳区中心、青岛流亭国际机场均在项目10 km范围内。

本设计旨在为城市社区提供一个居住与生活的全面解决方案。市政设施、社区服务及行政管理、多元化文体活动、多形态建筑产品、可持续发展的环保行为等被全面引入社区，使之形成自给自足的可持续发展的居住环境。

建筑风格力求具有创新性、独特性，塑造出沉稳、高雅的建筑形象。在符合本项目建设区域整体要求的同时，项目的建筑形象能够体现出"泰晤士小镇"的品牌理念。

This project is located in Xifu, Qingdao. It is next to the Laoshan scenic area. Jimo District, Chengyang District and Qingdao Liuting International Airport are within 10 km.

This design aims to provide a comprehensive solution for residence and life in urban communities. Municipal facilities, community services and administrative management, diversified cultural and sports activities, multi-form building products, sustainable environmental protection behaviors, etc. are fully introduced into the community to form a self-sufficient and sustainable living environment.

The architectural style strives to be innovative and unique to create a calm and elegant architectural image. While meeting the overall requirements of the construction area of the project, the architectural image of the project is enabled to reflect the brand concept of "Thames Town".

总用地面积：	82 hm²
总建筑面积：	1 287 000 m²
容积率：	1.54

Site area:	82 hm²
Gross floor area:	1,287,000 m²
F.A.R.:	1.54

青岛即墨南山果岭艺墅
Nanshan Golf Villas, Jimo, Qingdao

地点：中国，青岛
业主：青岛南山集团长基置业有限公司
设计范围：方案设计、扩初设计
设计时间：2010
建成时间：2013

Location: Qingdao, China
Client: Qingdao Nanshan Group Changji Real Estate Co., Ltd.
Design scope: Proposal Design, Design Development
Design time: 2010
Completion time: 2013

本项目位于青岛即墨，紧邻鳌山湾，是一座涵盖了国际博览中心、五星级酒店、企业会馆、游艇俱乐部、温泉疗养中心、滨海高尔夫、写字楼与高档居住区等多种功能的综合型滨海度假综合体。

主干道沿着南北方向的轴线展开，而支路则沿着东西方向的轴线展开。每一个建筑组团都被串联起来，在整个区域中形成简洁、高效的格局。综合体西侧可以欣赏到迷人的海景，而其东侧可以直接看到大片的高尔夫球场的景色。

The project is located in Jimo, Qingdao and adjoined to Aoshan Bay. It is a coastal resort complex with mixed-use facilities including a national exhibition centre, a five-star hotel, an enterprise clubhouse, a yacht club, a spa, a coastal golf course, office buildings, and upscale residential spaces.

The main avenue is oriented along the north-south axis while the secondary streets are oriented along the west-east axis. Each group of buildings is thoroughly connected, forming a simple and efficient distribution throughout the whole area. The west wing of the complex has a stunning view of the sea, while the east wing has a direct view of the vast golf course.

总用地面积：	1.4 hm²
总建筑面积：	42 000 m²
容积率：	0.3
住宅总户数：	107

Site area:	1.4 hm²
Gross floor area:	42,000 m²
F.A.R.:	0.3
Total amount of units:	107

Ruicheng Greenland Island, Yuhang, Hangzhou

地点：中国，杭州
业主：杭州瑞城房地产开发有限公司
设计范围：方案设计、扩初设计
设计时间：2003
建成时间：2009

Location: Hangzhou, China
Client: Hangzhou Ruicheng Real Estate Co., Ltd.
Design scope: Proposal Design, Design Development
Design time: 2003
Completion time: 2009

本项目位于杭州市，地块周边有西湖、径山等旅游资源。基地地形以丘陵为主，北邻凤兴路，南靠宋家山，东接良上公路，西接东西大道。基地的两个主出入口设置在良上公路和东西大道上。

设计依托项目独有的地貌，贯彻"有山有水"的居住小区及体育度假圣地的设计主题。

居住地块位于基地北部，依托自然山水形成生态型的居住小区，并因地形或房型的不同而分成若干个组团。旅游地块位于基地南部，引入以网球运动中心为主题，以会议中心度假村为载体的"健康运动休闲旅游区"概念，突出了休闲旅游项目的主题内容，提高了人气。居住区的封闭模式及旅游区的开放模式有效减少了相关设施的重复设置。环境设计上尽可能保持其自然的特色。

总用地面积：	20.3 hm²
总建筑面积：	270 000 m²
容积率：	1.2
住宅总户数：	1 842

The project is located in the city of Hangzhou, with West Lake and Mount Jing in the vicinity. The terrain of the base is dominated by hills, with Fengxing Road in the north, Mount Songjia in the south, Liangshang Highway in the east, and Dongxi Avenue in the west. The two main entrances of the site are located at Liangshang Highway and Dongxi Avenue.

Governed by a unique geographical condition, the design features a residential community with "water and mountain" and a sport resort as primary design themes.

The residential plot is located in the north of the site, relying on natural mountains and rivers to create an ecological residential community, which are divided into several groups due to different terrain or house types. The tourist plot is located in the south of the site, introducing the concept of "healthy sports and leisure tourism area" with the tennis sports centre as the theme and the conference centre resort as the carrier, highlighting the theme content of leisure tourism project and increasing popularity. The closed pattern of residential area and the open pattern of tourist area have effectively reduced the repetitive setting of related facilities. The environment is designed to maintain its natural features as much as possible.

Site area:	20.3 hm²
Gross floor area:	270,000 m²
F.A.R.:	1.2
Total amount of units:	1,842

无锡新吴千禧日式服务公寓
Millennium Japanese Style Service Apartment, Xinwu, Wuxi

地点：中国，无锡
业主：无锡鑫畅置业有限公司
设计范围：方案设计、扩初设计
设计时间：2005
建成时间：2008

Location: Wuxi, China
Client: Wuxi Xinchang Real Estate Development Co., Ltd.
Design scope: Proposal Design, Design Development
Design time: 2005
Completion time: 2008

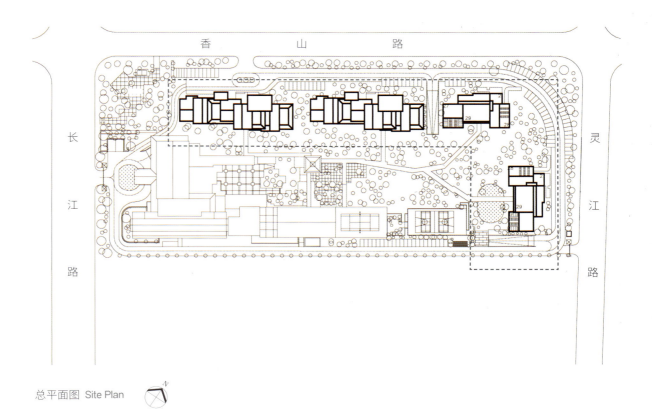

香　山　路

长江路

灵江路

总平面图 Site Plan

本项目位于无锡市。建筑围合出约40 m宽，150 m长的花园，满足了小区用户的需要。围合性的空间强化了项目宁静、高雅、舒适的感觉。北侧香山路设置有公寓区域的主出入口，与酒店的主出入口互不干扰。

各栋公寓楼通过中央景观空间中的景观回廊相连接，使业主充分享受高档社区的便利性与舒适性。同时，各主题景观空间在景观回廊中自然形成，体现了中式园林的特色。

本项目为建筑与室内的一体化设计服务。

The project is located in Wuxi. The architectures surround and shelter a 40 m wide and 150 m long garden to meet the requirements of the residents. This enclosure space emphasizes a feeling of calm, elegance and comfort. The entrance of the apartment area is located at Xiangshan Road on the north and is separated from the main entrance of the hotel.

Each apartment building is connected by the landscape corridors in the central landscape area, so that the owners can enjoy the convenience and comfort of the high-end community. Meanwhile, various themes of landscape spaces are naturally formed in the landscape corridors, embodying the characteristics of the Chinese gardens.

It is a completion of the full design service for both the architecture and interior of the project.

总用地面积：	3.8 hm^2
总建筑面积：	52 000 m^2
住宅总户数：	496

Site area:	3.8 hm^2
Gross floor area:	52,000 m^2
Total amount of units:	496

哈尔滨南岗松江新城
Songjiang New City, Nan'gang, Harbin

地点：中国，哈尔滨
业主：鲁商置业股份有限公司
设计范围：规划设计、方案设计、扩初设计
设计时间：2010
建成时间：2015

Location: Harbin, China
Client: Lushang Property Co., Ltd.
Design scope: Planning Design, Proposal Design, Design Development
Design time: 2010
Completion time: 2015

总平面图 Site Plan

本项目位于哈尔滨市中心南部的南岗区，为黑龙江省农业科学院旧址。该地区是哈尔滨市高校聚集地之一，具有浓郁的科技文化氛围，包含居住、商业、酒店、办公、公寓等业态，未来将建设成为功能完善、配套设施齐全、布局合理、特色突出的哈尔滨南部地区大型居住商业综合体。

建筑设计体现以居住为主，综合开发的思想，以提高建筑品质，丰富空间形式，创造良好的城市形象与居住环境。设计还提高了每个单元的视觉通透性，尽量减少南侧建筑对北侧建筑的遮挡。

This project is located in Nan'gang District—southern part of Harbin city centre. The former site is Heilongjiang Academy of Agricultural Sciences. This area is one of the gathering places of colleges and universities in Harbin. It has a strong technological and cultural atmosphere, involving residential, commercial, hotel, office, apartment and other business types. In the future, it will be built into a large-scale residential and commercial complex in the southern part of Harbin with complete functions, complete supporting facilities, reasonable layout and outstanding characteristics.

The architectural design embodies the idea of focusing on residence and comprehensive development, in order to improve the quality of the buildings, enrich the spatial form, and create a good urban image and living environment. The design also increases the visual transparency of each unit, and minimizes the shielding of the buildings on the south to the buildings on the north.

总用地面积:	40 hm²
总建筑面积:	966 124 m²
容积率:	2.43

Site area:	40 hm²
Gross floor area:	966,124 m²
F.A.R.:	2.43

大连甘井子大华锦绣华城
Dahua Jinxiu City, Ganjingzi, Dalian

地点：中国，大连
业主：大华集团大连置业有限公司
设计范围：规划设计、方案设计、扩初设计
设计时间：2011
建成时间：2014

Location: Dalian, China
Client: Dahua Group Dalian Real Estate Co., Ltd.
Design scope: Planning Design, Proposal Design, Design Development
Design time: 2011
Completion time: 2014

总平面图 Site Plan

本设计是在大华锦绣华城项目控制性详细规划的基础上，对项目F1区进行的建筑设计。

地块位于红凌路以东，向北至马栏广场，向南至大连理工大学北门。

该项目体现以居住为主，综合开发的思想，以提高建筑品质，丰富空间形式，创造良好的城市形象与居住环境。

The design is based on the regulatory plan of Dahua Jinxiu City project, to propose an architectural design for the Lot F1.

The lot is located on the east of Hongling Road, south of Malan Square and north of the north gate of Dalian University of Technology.

The project refers to the idea of focusing on residence and comprehensive development to create a high quality urban image and living environment by improving the architecture quality and enriching the spatial form.

总用地面积：	5.8 hm²
总建筑面积：	149000 m²
容积率：	2.29
住宅总户数：	1 390
绿化率：	40.20%

Site area:	5.8 hm²
Gross floor area:	149,000 m²
F.A.R.:	2.29
Total amount of units:	1,390
Rate of greening:	40.20%

INTERIOR DESIGN

室内设计

无锡新吴无锡千禧大酒店
Wuxi Millennium Hotel, Xinwu, Wuxi

地点：中国，无锡
业主：无锡鑫畅置业有限公司
设计范围：室内设计
设计时间：2006
建成时间：2009

Location: Wuxi, China
Client: Wuxi Xinchang Real Estate Development Co., Ltd.
Design Scope: Interior Design
Design time: 2006
Completion time: 2009

酒店的室内设计延续了酒店建筑的日式风格，保持了建筑设计的延续性。

酒店地上为二十二层，地下为一层，三层至二十二层是酒店客房。地下一层、一层、二层是公共区域。地下一层布置了全日餐厅、超市和停车场，一层有大堂、大堂休息吧、日式餐厅、游泳池、浴场和健身房，二层是宴会厅、会议室、中餐厅、水疗等功能空间。

大堂的室内设计采用了半开放式的设计，将室外景观作为私密空间和公共空间的隔墙。设计引入人工瀑布作为大堂的背景墙。大堂的下沉地面使瀑布成为大堂的重要元素，并从视觉上将主入口和水景联系起来。与景观元素相结合的理念贯穿在室内公共区域的设计中。弧形顶的设计替代了常见的平顶设计。设计在弧形顶上使用传统日式图案来丰富弧形顶的造型。弧形顶的图案又与背景墙的图案以和谐的方式形成呼应。

全日餐厅在地下一层，靠近室外景观瀑布的一侧采用了玻璃材质。将室外景观引入室内的设计概念与大堂的设计概念一致。尽管餐厅设置在地下一层，但这里的就餐氛围更像是在地上一层或二层就餐的感觉。延伸出去的平台增强了人们对瀑布的亲近感，将人与自然和谐地联系起来。

总统套房的室内设计体现出富丽堂皇的感觉，采用了欧式风格进行装饰，金箔、精致的线条和真皮家具等装饰元素都有助于奢华氛围的营造。

酒店客房的室内也采用了日式风格设计：墙面涂有白色乳胶漆，窗户周围带有木框线条。酒店客房同时也配有日式风格的家具。

The interior finishing of the hotel follows the Japanese style exterior of the hotel, and retains the continuity of the exterior.

The hotel has 22 floors above ground and one floor underground. The hotel rooms start from the 3rd floor to the 22nd floor. The underground floor, the ground floor and the first floor are public spaces. The underground floor accommodates a 24-hour restaurant, a supermarket and a parking lot. The ground floor includes a lobby, a lounge bar, a Japanese style restaurant, a swimming pool, a bathhouse and a gym. Banquet halls, conference rooms, a Chinese style restaurant and a spa are located on the first floor.

The interior of the lobby adopts the semi-open design that incorporates partitions for both intimate and public spaces by using the exterior landscape. The design introduces an artificial waterfall as the backdrop for the lobby. A sunken floor for the lobby makes the waterfall a prominent element and visually connects the main entrance and the waterscape. The concept incorporating landscape elements is used throughout the public interior design. An arc-formed roof replaces the commonly used flat design. The design implements the traditional Japanese patterns to enrich the shape of the ceiling. The patterns of the ceiling respond to the decorative patterns of the walls in a harmonious way.

A 24-hour restaurant is located on the underground floor. One side of the restaurant that is adjacent to the waterfall is designed in glass. The design concept of introducing the exterior landscape into the interior is the same as the lobby. Although the restaurant is located underground, the dining atmosphere is more like dining on the ground floor or the first floor. The extended terrace increases the immersive feeling of the waterfall, integrating people and nature in a harmonious way.

The luxurious European style interior design for the Presidential Suites creates a feeling of grandeur, decorative elements such as gold foil leafs, delicate lines, and leather furniture contribute to this atmosphere of luxury.

The interior of the hotel rooms are also designed according to the Japanese style, where the walls are painted white with wooden frames complementing the window openings. The hotel rooms are completed with Japanese-styled furnishing.

上海松江达安圣芭芭花园河谷3号
Da'an St. Babara Valley No.3, Songjiang, Shanghai

地点：中国，上海
业主：上海达安泰豪置业有限公司
设计范围：室内设计
设计时间：2007
建成时间：2009

Location: Shanghai, China
Client: Shanghai Da'an Taihao Real Estate Co., Ltd.
Design scope: Interior Design
Design time: 2007
Completion time: 2009

本项目的室内设计延续了室外的空间感，并且进一步强化了室内空间的连续性，增强了室内空间感。开放的地下室空间是庭院的延伸。设计包括对卧室空间到卫浴空间的延伸部分，以及卧室空间和室外景观之间的连接。

河谷3号共有五种房型，其室内设计采用了地中海风格并与外立面相互呼应。室内设计保持了特定的空间分隔，并与建筑设计保持一致。装饰立面和地面在材料上有所区分。

每种房型的地下室都设有天井，可以把阳光引入室内。这种设计彻底颠覆了人们对地下空间的通常印象。

卫生间与主卧室相结合。在一些房型中，设计使用透明玻璃隔断将卫生间和卧室分隔开，制造出一种二者完全隔开的错觉。床或浴缸的上方开有天窗，使房间的这一区域特别明亮。夜晚，天窗使人可以在室内看到月亮和星星，给人带来惊叹之感。这种大胆而有创意，温情而浪漫的设计，不仅节省了空间，创造了价值，而且增添了一种神秘感。

马赛克在房型的设计中也有大量的运用，主要以浅灰色或白色马赛克作为背景，整体效果带来丰富的视觉享受。颜色的大量使用表示不同的功能和用途。蓝色马赛克表达一种存在感，暖橘色马赛克表示这一区域为休息空间。所有房型的面积都比较小，因此设计采用了大量的玻璃隔断来加强空间感，并从视觉上扩大空间感。

The interior design of the project retains the same spatial architectural atmosphere as the exterior. The design emphasizes the continuity of the internal space as to amplify the effect of spatial conditions. The open underground level is an extension of the garden. The design includes the extension of the bedroom space into the bathroom space as well as the connection between the bedroom space with the exterior landscape.

Valley No.3 has a total of five different unit models. The interior design of the five models is in Mediterranean style and responds to the exterior facades. The design of the interior retains specific spatial separation and is in accordance with the architectural design. The decorative facade and the ground are separated by materials.

All the unit models have an open courtyard at the underground level to introduce sunlight into the building. This design consideration completely transforms people's notion of the underground space.

The bathroom is integrated in the master bedroom. In some unit models, the design makes use of transparent glass to set the bathroom apart from the bedroom, creating an illusion of a complete separation. Some skylights have been designed above the bed or the bathtub to make these parts of the room particularly bright. In the evening, the skylights give a sense of awe from seeing the moon and the stars from the interior. This bold, creative, loving and romantic design saves the space, creates a value and contributes to an increased sense of mystery.

The unit models are extensively decorated with mosaics on a background of light gray and white mosaics, the result is an experience full of visual enjoyment. Colours have been used extensively to denote functions and uses. The blue mosaic expresses a presence and the warm orange coloured mosaic announces an area intended for relaxation. The unit models are compact in size, so the design makes extensive use of glass walls to enhance and visually enlarge the spatial experience.

上海静安加拿大KFS国际建筑师事务所办公楼
Office of KFS Architects International Inc. Canada, Jing'an, Shanghai

地点：中国，上海
业主：加拿大KFS国际建筑师事务所
设计范围：室内设计
设计时间 Design: 2005
建成时间 Completion: 2008

Location: Shanghai, China
Client: KFS Architects International Inc.
Design scope: Interior Design
Design time: 2005
Completion time: 2008

上海静安加拿大KFS国际建筑师事务所办公楼
Office of KFS Architects International Inc. Canada, Jing'an, Shanghai

整个办公室的设计主题简洁、现代、优雅。玻璃隔断的大量运用，消除了室内和室外的空间界限，把室外的景观引入室内来。

接待处一进门就可以看到四只镇室之兽整齐地摆放在墙的一侧，与浅色的室内墙面形成对比。接待处的照明由四盏红灯笼组成，预示着办公室的生意红红火火，且充满了传统东方的个性。

房间门的中间都是透明玻璃，玻璃背后有罗马帘，它控制了屋内办公空间和屋外花园景观的关系。把罗马帘拉上去时，屋内与屋外的关系不言而喻；把罗马帘拉下来时，屋内就变成了一个私密的空间。

大办公室采用开放式设计，给工作人员提供了一个愉悦的工作环境。

楼与楼之间的中庭挑空了三层，并采用半封闭式的设计。前后门都采用了玻璃材质，将中庭和室外空间联系起来，使有点狭小的空间没有区隔。前后花园的景色作为一个整体被引入中庭，体现了中式传统庭院的概念。

The theme of the office design is simple, modern and elegant. Extensive use of glass in partitioning spaces eliminates defined boundaries between interior and exterior space. The exterior landscape is introduced into the interior.

Upon entering the reception area, four mythological creatures are placed in a row in contrast to the neutral interior wall behind. The interior lighting for the reception area are composed of four red lanterns that indicate the conduct of brisk business and the full traditional eastern personality of the office.

The middle of the doors are transparent glass. The Roman blinds, installed behind the glass, control the relation between the interior offices and the exterior landscaped gardens. When the Roman blinds are in the open position the connection to the exterior is self-evident. When the Roman blinds are down, the office is transformed into an intimate space.

The open-style design of large offices provides a pleasing working environment for the staff.

The three-storey atrium between each building has semi-enclosed roofs. The rear door and the door to the street are fully glazed to link the atrium with the exterior space, making the moderate-sized room partition free. The atrium exhibits the traditional Chinese courtyard-philosophy of integrating the front garden and back garden as a single entity.

项目年表 CHRONOLOGY